The
Great
Displacement

———

CLIMATE CHANGE
AND THE NEXT
AMERICAN MIGRATION

———

Jake Bittle

SIMON & SCHUSTER

NEW YORK   LONDON   TORONTO   SYDNEY   NEW DELHI

Simon & Schuster
1230 Avenue of the Americas
New York, NY 10020

First Simon & Schuster hardcover edition February 2023

SIMON & SCHUSTER and colophon are registered trademarks of Simon & Schuster, Inc.

For information about special discounts for bulk purchases, please contact Simon & Schuster
Special Sales at 1-866-506-1949 or business@simonandschuster.com.

The Simon & Schuster Speakers Bureau can bring authors to your live event. For more
information or to book an event, contact the Simon & Schuster Speakers Bureau at
1-866-248-3049 or visit our website at www.simonspeakers.com.

Interior design by Alexis Minieri

Manufactured in the United States of America

3 5 7 9 10 8 6 4

Library of Congress Cataloging-in-Publication Data has been applied for.

ISBN 978-1-9821-7825-3
ISBN 978-1-9821-7827-7 (ebook)

To my mother and father

Contents

You cannot put a Fire out—
A Thing that can ignite
Can go, itself, without a Fan—
Upon the slowest Night—

You cannot fold a Flood—
And put it in a Drawer—
Because the Winds would find it out—
And tell your Cedar Floor—

Emily Dickinson

The
Great
Displacement

Introduction

The town had been there for a century and a half. Then one evening, in the summer of 2021, it disappeared.

Greenville was a mountain hamlet of about a thousand people, nestled in the mountainous wilderness of northeast California. The town had sprung up as a mining outpost during the gold rush, and buildings from that era still dotted its main street. The local economy was sluggish, and most residents were far from wealthy, but the community was close-knit; as one resident later recalled, "We'd go to the local grocery store and sometimes it would take over an hour to get out, just to run in and get a gallon of milk or something, because we knew everybody." Some residents had moved to Greenville from the Bay Area in search of cheap housing, while others relished the town's distance from urban sprawl—a writer from Greenville would later recall a social tapestry made up of "retirees, hippies, bikers, rednecks, ranchers, [and] cowboys." Inside the valley, insulated from the mountains by a screen of trees, it was easy to forget the rest of the world.

On the evening of August 4, 2021, one of the largest wildfires in US history breached the valley surrounding Greenville. The Dixie Fire had been smoldering for weeks nearby, but a sudden wind from the south had sent it spiraling to the northeast. It moved through the valley "like a blowtorch," as one resident put it, destroying three-quarters of the town's buildings in a matter of minutes. Houses erupted into fluttering tufts of flame;

cars shriveled up like dried flowers; light poles and stop signs doubled over; trees dissolved into yellow air. The fire obliterated the library, the pharmacy, the baseball field where the Little League teams played, and the streets that hosted the annual "Gold Diggers" mining festival; it incinerated boxes full of family photos, heirloom rifles, children's toys, and workaday silverware sets. A hundred and seventy years of history and memory evaporated into soot and haze.

The disaster did not end once the fire burned out. It took months for the federal government to arrive in Greenville with emergency trailers that could shelter displaced residents. In the meantime, the refugees from the town scattered across the state to find temporary housing, forced into a wrenching exile from their homes. Greenville's working-class residents had come to the town because it was affordable, and now they struggled to find alternative shelter elsewhere in the state: California's housing shortage had jacked rents up to unprecedented levels even in places that had seen no wildfire damage, and many victims had to search for months before they found apartments they could afford. Some were priced out of California altogether, while others left the state by choice, terrified of living through another fire season—a few of the victims had moved to Greenville from the town of Paradise after the Camp Fire three years earlier, only to find themselves burned out again. Meanwhile, many who wanted to return to Greenville found that they lacked the means to do so. In the years before the fire, insurance companies had started to pull out of the town, telling residents the area was too risky to insure, and many victims who had lost their homes in the fire did not have sufficient insurance coverage to rebuild. They moved to nearby towns and cities instead, ten or twenty miles away, and found themselves slipping into new lives. The few hundred residents who did return found the town in a state of permanent limbo: the scarred landscape around them was haunting and hostile, frozen in the moments after the fire. The natural beauty of Greenville's secluded valley, the affordable housing stock that had attracted artists and eccentrics, the close social ties that had bound the community together—everything that made Greenville what it was had vanished.

The destruction of this quirky mining town and the displacement of its residents function as a parable for what the next century of climate change will bring to the rest of the country. The summer of the Dixie Fire saw a series of climate calamities strike almost every corner of the United States—there was another historic fire near Lake Tahoe, a monstrous hurricane in Louisiana, a millennium-scale drought across the Southwest. Thousands of people lost their houses to a tide of storm surge, or watched their communities go up in smoke, or suffered beneath lethal temperatures. By the end of the year, one in three Americans had experienced a weather disaster of some kind.

For a long time, climate change was something to be discussed in abstract terms, something that existed in the future tense. That is no longer the case. Each passing year brings disasters that disfigure new parts of the United States, and these disasters alter the course of human lives, pushing people from one place to another, destroying old communities and forcing new ones to emerge.

This book tells the firsthand stories of people whose lives have already been touched by climate change, who have already lost their homes and their histories to a crisis that millions more of us will soon confront. It draws on hundreds of original interviews and thousands of pages of research to show how climate disasters expose fundamental flaws in where and how we have chosen to build our communities. It also illustrates how government disaster policy and the private housing market combine to push people away from the riskiest places in the aftermath of floods and fires, creating a cycle of displacement and relocation. But this book is also a portrait of a generation of domestic climate migrants, one that is growing larger every year. These people live in every corner of the country, from the waterlogged streets of Miami to the parched cotton fields of Arizona. They run the gamut from minimum-wage workers to millionaires, from liberals in big coastal cities to entrenched small-town conservatives. Their stories range from the tragic to the comic and from the inspiring to the infuriating.

Indeed, there is only one thing they all have in common: they are moving.

———

The title of this book is an oblique reference to the Great Migration, the largest single migration event in American history. From the 1920s to the 1970s, more than six million Black people left the South and moved to northern cities like New York and Chicago, fleeing an economic and humanitarian crisis. They left to escape the yoke of Jim Crow, the rampant flooding along the Mississippi, and the economic stagnation of the former plantation states, and flocked to the north for good-paying industrial jobs and the promise of racial tolerance. There have been other nationwide migrations since, from the Dust Bowl exodus to the Sun Belt boom, and many of these migrations have been driven by crises and upheavals, but none anywhere near so consequential as the Great Migration: the Harlem Renaissance, the civil rights movement, and the rise of hip-hop and blues can all be traced back to this seismic shift.

By the end of the century, climate change will displace more people in the United States than moved during the Great Migration, uprooting millions of people in every region of the country. In recent years, politicians and academics have begun to use the term "climate migration," but "migration" is not quite the right word to describe what is happening in the aftermath of disasters like the Dixie Fire. The word implies an intentional, one-directional action, but the movement on the ground is more diffuse. It is beginning everywhere, in fits and starts, and enfolding people of all races and classes. In aggregate, the movement looks less like an arrow shooting through the air and more like water churning in a pot as it reaches a boil. Hence the use in this book of the term "displacement," which conveys the reality: these movements will be unpredictable, chaotic, and life-changing.

This churn of displacement is occurring amid many other kinds of upheaval. Disasters like the Dixie Fire are not happening in a vacuum—they are happening to a society that is lopsided and unequal, shaped by political and economic forces. The blaze that destroyed Greenville may have ignited due to a climate-induced drought, but the town's exodus was the product of an underfunded disaster relief system, a dire affordable housing shortage, and a broken insurance market. These same factors are fueling displacement in other parts of the country, after other kinds of disaster: climate change is applying stress to an already brittle social and economic order, widening cracks that have been there the whole time.

At the most fundamental level, displacement begins when climate change makes it either too risky or too expensive for people to stay somewhere. The disasters discussed in this book bear little resemblance to each other on the surface, but they all exert pressure on governments and private markets, whether through the financial costs of rebuilding or the strain of allocating scarce resources. As this pressure builds, it starts to push people around, changing where they can live or where they want to live. Sometimes this looks like the government paying residents of flood-prone areas to leave their homes; sometimes it looks like fire victims getting priced out of an unaffordable state; other times it looks like fishermen going broke as the wetlands around them erode. It may seem reductive to think about a planetary crisis in terms of financial risk rather than human lives, but that is how most people in this country will experience it—through the loss of their most valuable assets, or the elimination of their job, or a shift in where they can afford to live.

The burden of this shift will not fall on everyone equally—indeed, displacement will create new cleavages between the rich and poor, the privileged and the marginalized. If the government can only spend so much money on flood walls, it might choose to protect wealthier communities with more robust tax bases; if a thousand fire victims scramble for two hundred vacant apartments, the richest two hundred renters are more likely to end up with roofs over their heads. These disasters will serve as reckonings for a society that has attempted to tame the forces of nature while leaving the provision of shelter to the whims of the market.

As more people leave their homes, patterns of displacement will emerge. The trend will begin in small towns in remote places, but over time the instability will spread to major cities and entire regions. As disasters continue to pummel our housing markets, public and private powers will push more people out of vulnerable areas, and escalating fear of danger will spur more to move of their own volition. The result will be a shambling retreat from mountain ranges and flood-prone riverbeds, back from the oceans, and out of the desert. It will take decades for these movements to coalesce, but once they do, they will reshape the demographic geography of the United States.

Even today, in towns across the country, the contours of this new national future are starting to emerge.

—

Among people who live in more temperate locales, there is often a tendency to gawk at people who choose to live in places that appear vulnerable to disaster, like New Orleans or the fire-prone Sierra Nevada range. In the aftermath of a hurricane or wildfire, it's common to hear the question "Why don't they just move?" The past few years have shown that way of thinking to be utterly misguided, as disasters have struck areas that many people considered insulated from climate risk. The summer of 2021 saw a lethal "heat dome" pass over the Pacific Northwest, killing hundreds of people. The following month, a series of devastating floods crashed through the mountains of middle Tennessee, wiping away whole towns and sweeping children away in rushing water. The month after that, the remnants of Hurricane Ida killed more people in New York and New Jersey than the storm itself did in Louisiana. That same year saw a deep freeze in Texas, a tornado squall in Kentucky, and an urban wildfire in the Denver suburbs. The climate crisis is coming for everyone.

Even so, not all risks are created equal. There are some places in the

United States that are far more vulnerable than others, places where climate risk is not occasional but existential. The towns and cities profiled in this book are among the first places in the United States where climate displacement has begun to unfold, but their stories are representative of broader trends taking hold in other places across the country. They illustrate how displacement begins at the local level in the aftermath of a disaster and widens over time to pull thousands of people into the current.

The first three chapters examine the three major forces that create climate displacement. The first and most obvious force is the ever-increasing severity of weather disasters, so we begin in the Florida Keys, an archipelago of islands pushed into a state of permanent decline by rising water. The second force is government policy; for this chapter we move to a small town in rural North Carolina, where a government buyout program emptied a historic Black neighborhood. The third and most chaotic force is the private housing market; this chapter takes place in Santa Rosa, California, where a housing shortage upended the social order after devastating wildfires.

The next two chapters explore how these forces combine to create the large-scale, long-term displacement that will characterize the next century. The fourth chapter takes place on the coast of Louisiana, where coastal erosion has threatened the survival of an Indigenous fishing village; the fifth moves west along the Gulf of Mexico to Houston, where a series of floods has ravaged the city's urban fabric and torn its neighborhoods apart along racial and economic lines.

The final three chapters pan out to the regional migrations that are already taking shape. The sixth chapter recounts the emerging water crisis in central Arizona, which has destroyed the region's cotton-farming industry and has jeopardized the future growth of the Phoenix suburbs. The seventh moves east to the coastal city of Norfolk, Virginia, where rising sea levels have destabilized the flood insurance system and set the stage for a housing crash of epic proportions. The final chapter takes up the thorny question of where people displaced by climate change will land, surveying the country to chart the trajectory of the next century's migration.

The protagonists of this story are the thousands of people who have already left their homes and communities behind; in doing so, these people have become some of the first Americans to experience what the writer Sonia Shah calls "life on the move," a strange and uprooted existence in a turbulent world. This book tries to illuminate the climatological forces that push climate migrants out of their homes and the political and economic structures that determine where they end up, but it also tries to memorialize the many places that climate change will force us to abandon. Even in a world that is only one degree warmer, floods and fires are already destroying fragments of history and culture, and these losses are impossible to quantify. The forces that shape climate displacement may be vast and impersonal, but the experience of leaving home is always intimate, and every displacement story is also a story of personal bereavement. What gets lost when a fire consumes a place like Greenville are not just houses and streets, but floors that witnessed children's first steps, kitchens that hosted nighttime arguments, cul-de-sacs that signaled the end of evening commutes. When a community disappears, so does a map that orients us in the world.

On the other side of this loss, there is an opportunity. The sheer scale of displacement over the coming century forces us to reconsider our relationship to the places we call home—not only the houses we inhabit but also the land we occupy. Why have we built so many homes in places that are vulnerable to floods and fires? Why are our housing markets so unfair and unaffordable? Who gets to decide where we should and should not live? The way we answer these questions will determine what life will look like for future generations, who will have to live with the consequences of further warming. One thing is certain: The status quo is not working. Millions of people live at perennial risk of losing their homes to climate disasters. When people do lose their homes, many of them struggle to find new ones, and often find themselves in places that are worse off and more vulnerable than the ones they left behind. The government agencies and private industries designed to help them recover often push them toward further instability. There are ways to fix this broken system, to ensure that our communities are resilient to climate disasters and that our society can guarantee shelter for everyone, but first we must acknowledge how large the crisis of displacement has already become.

It has become commonplace for journalists to call climate change "the story of the century," and it's true that no phenomenon will exert a greater influence on American life over the next hundred years. This book is an attempt to tell the first act of that saga, to follow the displacement and upheaval that is already underway in the United States. The rest of the story is yet to be written.

CHAPTER ONE

The End of the Earth

———

CLIMATE CHANGE

AND THE AGE OF

PERMANENT DISASTER

———

Big Pine Key, Florida

I.

The Big One

In late June of 2017, Jen DeMaria drove down to Key West for Tropical Fruit Fiesta, a celebration of exotic fruit hosted by her friend Patrick Garvey. She and Patrick both lived on Big Pine Key, a hardscrabble island about half an hour to the east, part of a chain of islands called the Florida Keys. Big Pine is known as a refuge for misfits, and Jen and Patrick owned businesses that matched the island's reputation: Jen and her fiancé ran a vegan beachfront bed-and-breakfast, and Patrick tended a tropical fruit grove that had once been the property of a local hermit.

To get from Big Pine to Key West, Jen crossed a succession of barely there tropical islands dangling off the southern tip of Florida, each thick with mangroves and overrun by lizards, roosters, and endangered deer. Surrounding her on both sides was an ocean colored an almost outrageous shade of swimming pool blue, capped by billowing clouds and a crystalline sky. When she arrived at the Fiesta, she found Patrick as happy as she'd ever seen him. He was handing out pamphlets about his fruit propagation efforts, chatting with fellow horticulture experts, and serving up exotic juices with help from his friend Ukulele Tim. His wife, Angelica, and their three-year-old twin daughters were there, too, along with their dog, Bella, the whole family sweating in the afternoon heat.

Patrick had moved to the Keys from Prince Edward Island in Canada, certain that he was searching for something, but uncertain what that something was—at any rate, he wasn't thinking about tropical fruits. He spent a few years as an investigator for the state's Department of Children and Families before someone told him about a destitute fruit grove on Big Pine. The grove had been the property of one Adolf Grimal, a reclusive retiree who had smuggled countless rare fruits into the Keys in an attempt to build a global nursery. After Grimal died in 1997, the property fell apart, and when Patrick first saw the grove it was serving as an open-air drug den. With a single-minded devotion, Patrick tracked down Grimal's family and bought the plot from the old man's surviving kin, then emptied his savings to restore it. After years of pulling roots and chopping wood, his work had paid off: the grove had a healthy population of jackfruits, longans, soursops, and other fruits that existed almost nowhere else in the continental United States. What had at first seemed to Patrick's neighbors like a futile vanity project had become a bona fide community resource—in addition to a nursery, the grove was also a living laboratory for horticultural education and a gathering place for local hippies. It was one more curiosity the residents of Big Pine Key could be proud to call their own.

That day in June of 2017 turned out to be the high point of Patrick's life. The following month, Angelica and the twins went to Brazil to spend the fall with Angelica's family, leaving Patrick alone to work the grove. A few weeks later, in late August, meteorologists began tracking a tropical storm system called Irma, which churned across the Atlantic with record-breaking intensity ahead of Labor Day weekend.

The Florida Keys sit a mere three feet above sea level on average, and their pendulous position in the Gulf of Mexico makes them a prime target for tropical storms. But island residents—known in local lingo as "conchs"—often adopt a defiant posture toward hurricanes, and pride themselves on sticking it out rather than evacuating to the mainland. Irma was different: the storm broke multiple records even before it hit land, maintaining Category 5 strength for sixty straight hours while still in the Atlantic. Even before the models had determined where the storm would hit, nearly everyone on the Keys packed up and left, booking hotels wherever they could find them in Tallahassee, Jacksonville, Georgia, and North Carolina. More than six million other Floridians did the same thing, pouring out of the state in bumper-to-bumper traffic that filled both sides of the highway.

At first, Patrick, Jen, and Jen's fiancé, Harry, planned to stick it out on Big Pine with two of their other friends, but at the last moment those two friends ditched them and disappeared from the Keys overnight. By the time Jen started getting scared, it was too late to evacuate. Every gas station from Key West to Miami and beyond had run out of fuel. She and Harry decided to wait out the storm with Patrick in a beach house down the street from their bed-and-breakfast.

On September 9, after Hurricane Irma made landfall in Cuba, it veered sharply to the north and pushed toward the Keys. Like most tropical storms, Irma had weakened as it passed over land, dropping from a Category 5 down to a Category 2, but the storm now regained power as it passed over the warm waters of the Straits of Florida, strengthening into a Category 4 storm with peak winds of more than 130 miles per hour. A storm that destructive had not made landfall on the Keys in more than eighty years, and almost none of the buildings on the archipelago were built to withstand it. Multiple people later recalled watching emergency officials on television tell anyone still in the Keys to write their Social Security numbers on their arms so rescue workers could identify them if the worst came to pass.

With hours to go before the storm arrived, Patrick, Jen, and Harry realized it wasn't safe for them to stay at the beach house. They left the house and made for the Sugarloaf School, a designated "refuge of last resort" a few islands over. There they found a scene that resembled a modern-day Noah's ark: hundreds of people had gathered in the school cafeteria along with pet cats, dogs, parrots, and fish, all having abandoned their homes and houseboats and campers. There wasn't a backup generator—someone from the county had arrived earlier and hauled it off.

Irma arrived at around three o'clock the next morning, slicing through the midsection of the Keys with eight feet of storm surge and winds that exceeded 150 miles per hour. The refugees at the Sugarloaf School sat on the floor of the cafeteria and listened as the roar of wind and water became first cacophonous, then deafening, drowning out all conversation. The lights shut off and the wind screamed around the building for what seemed like hours.

Then, all at once, the whole world seemed to go quiet: the tranquil eye of the storm was passing overhead. Bleary-eyed from a sleepless night, Patrick and Jen stepped outside into the early-morning light with their fellow refugees. Looking around the parking lot, they discovered that the wind had thrown palm trees onto many of their cars. Looking beyond the lot, they saw that the rest of the island appeared to have vanished into the ocean. The group rushed back inside before the opposite eye wall arrived and the gale started again.

As the storm wore itself out, the refugees at the Sugarloaf School realized they had been thrust back into the Stone Age. There was no power or phone service anywhere on the Keys, which meant there was no way for the holdouts to communicate with the emergency officials who would soon descend on the islands. They were on their own.

In the early afternoon, once the worst winds had passed, a few people in the cafeteria decided to take their chances driving back home, where they had backup generators and emergency radios. "If I'm not back in an hour," one man said, "that means I made it." Patrick, Jen, and Harry chose to risk it as well. Patrick took a friend in his truck and drove west to drop the friend off; Jen and Harry, meanwhile, headed east to Big Pine. They told Patrick to meet them at the bed-and-breakfast.

As Jen and Harry drove toward Big Pine, windshield wipers beating against the rain, they saw destruction that felt like something out of a dream—cars and sailboats flipped and splintered in the roadway and around bridge bases, motor homes and refrigerators cast around like bowling pins, trees uprooted and slammed through the windows of houses. At last they came to the turnoff for their street, Long Beach Drive, at the foot of which sat their bed-and-breakfast. But as Harry turned off the main road, Jen let out a gasp. There did not seem to be a Long Beach Drive. There were no homes, no street signs, nothing recognizable—only a vast chaos of debris and rushing water, a deluge of storm surge and detritus slipping back into the ocean. They gave up and reversed out to the main highway.

Patrick was waiting for them there. One of his neighbors, he told them, had given him a key to a camper where the three of them could stay for a few nights, at least until the electricity came back on. As she and Harry followed Patrick across Big Pine, Jen stared out the passenger window in shock. The damage on the island was so extensive that she had no idea where she was. The eye of the storm had missed Big Pine, passing over the Sugarloaf School a few miles west, but that wasn't a good thing; because hurricanes rotate counterclockwise, Big Pine had been hit by the "dirty side," which meant a devastating combination of storm surge and extreme wind. The storm had stripped the beaches of sand, flattened the inland forest, and carried away entire neighborhoods. Fallen trees had rendered many roads all but impassable, which meant there was no way to know who was out there and who needed help. When the three of them arrived at the camper, which had somehow been spared by the storm, Jen thought that the area seemed familiar, but she couldn't place it.

Toward the end of the day, when the sky started to clear, Patrick asked Jen to go out with him for a walk. They stepped out of the camper and walked through the yard, passing over a downed fence line and into the property next door. At first Jen thought she was standing in an abandoned lot—there were tree stumps scattered across the grass at random intervals, wood and metal strewn around like bird feed. About a hundred feet away she saw a single tree left standing, denuded of leaves. Patrick was silent as they surveyed the damage.

Jen, still dazed from the shock of the storm, looked up into Patrick's eyes.

"Where are we?" she asked.

He looked at her, his eyes blank.

"Do you really not know?" he said.

"No."

"Jen," he said, "we're in the grove."

She watched as Patrick turned and walked away, moving step by tentative step through the ruin of his life's work.

—

The following day, Jen and Harry told Patrick they were going back down to Long Beach Drive. They needed to get to work fixing the bed-and-breakfast, where storm surge had punched out the entire first floor and poured dead fish through second-floor windows. They were worried about Patrick, and they wanted him to come with them. There was an intact home near the bed-and-breakfast where they could squat for a few days until temporary trailers arrived, and the waves had carried off much of the debris, which at least made the road passable. Given the state of the rest of the island, it was just about the best place they could be.

Patrick told Jen no—he was going to stay near the grove.

"I have to be there," he told her. "I have to fix it."

Jen was scared for Patrick, and she thought about trying to talk him out of it. He wouldn't be able to stay at the safe house for more than a few nights, and the storm had flooded his own camper and wrecked two of his cars; going back to the grove would mean living in stifling heat without running water or air-conditioning. To make matters worse, the area around the grove was abandoned, and there would be nobody nearby who could help him with food or supplies. Patrick was undeterred—the grove was all he had, and he wasn't going to let it slip away.

If the Sugarloaf School had felt reminiscent of Noah's ark, Patrick's next few months in the grove recalled the story of Job, the saga of a man tested by a vengeful God. Angelica couldn't bring the twins back from Brazil, which left Patrick and his dog, Bella, alone on an island that felt like a ghost town. Every morning he walked out into the grove to confront a devastation that was almost incomprehensible, and every night he fell asleep alone in his battered, leaking camper, surrounded by a haze of mosquitoes.

This living situation, in addition to being intolerable, also ran afoul of the law. The Federal Emergency Management Agency, the government agency in charge of disaster recovery, had long maintained a so-called 50 percent rule for properties in flood-prone areas: if a property incurred damages worth more than half its value, the owner had to rebuild the property from the ground up, elevating it above flood level. This meant Patrick was not legally allowed to repair his camper now that the storm had destroyed it. According to the letter of the law, he had to build a new home on elevated stilts, something he didn't have the money to do. The FEMA policy was intended to help make housing safer in the aftermath of hurricanes, but it often rendered victims homeless if they couldn't afford to rebuild or move elsewhere, and in Irma's aftermath it turned many survivors into squatters. In the weeks after the storm, county officials roved back and forth across the Keys, evaluating the damage to individual properties and affixing orange or red tags to the houses that had sustained major damage. If the officials caught a homeowner trying to repair such a home, they could slap that person with a hefty fine.

The grove, meanwhile, had lost much more than half its value. The few trees that hadn't blown away had been sickened at the root by a surge of salt water, wiping out what Grimal had spent decades building and Patrick had spent half a decade rebuilding. Only one breadfruit tree had emerged healthy enough to survive on its own, and the rest had to be replanted and nursed back to health. The grove was in worse shape than it had been when Patrick found it, but this time he only had a few months to restore it; if he couldn't anchor new plants by the time the dry winter season arrived, the nursery would perish by the spring.

Even in those first weeks, Patrick had a vision for how to rebuild the grove. He couldn't save all of Grimal's rare plants, but he could use the seeds from the sole surviving breadfruit tree to create a whole colony of such trees, and could sell the breadfruits to restaurants, nurseries, and amateur horticulturalists. Before he could do that, though, he had to fix the property, and that meant hours upon hours of solitary labor—hauling out dead branches and heaps of storm-borne plastic, rebuilding hutches and nursery buildings, digging and replanting ruined stretches of soil. A few locals offered to help him raise money, and a friend flew down from Washington, DC, to help him for a few weeks, but other than that, he was on his own. Everyone else on Big Pine had their own crises to deal with, and the grove was his burden to bear.

As he pieced the grove back together, Patrick tried to remember why it had attracted him so much in the first place. He had always thought of himself as a storyteller, and the grove's mythical status had afforded him plenty of material to spin a compelling yarn. At one point, for instance, Grimal's family had told Patrick that the old man had smuggled the grove's lychee tree out of China, stealing a sapling from an imperial palace and sneaking away under cover of darkness, and Patrick had woven the tale into the spiel he gave during his tours of the grove. The story might not have been entirely factual, but it expressed something about the magic of the place the old man had built.

Soon that magic was all Patrick had left. By the new year he had begun to undo the damage the storm had done to the grove, but there was no undoing the damage it had done to his life. His family was nowhere to be found, he couldn't go back to Canada for fear the grove would collapse in his absence, and his friends in the Keys had drifted apart during the harrowing aftermath of the storm. Angelica and the twins came back for a few weeks, then left again, then came back for a few weeks around the holidays.

On the second visit, Angelica told Patrick that she wanted to separate—"I can't do this anymore," she said. Not long after, she and the twins departed for Brazil again.

Patrick soon ran out of money for repairs. He started scrounging for landscaping and maintenance jobs to stay solvent. As the rainy season ended and he struggled to coax the grove through the winter, his loneliness became all-consuming. It drowned out the progress he had made on the breadfruit and erased whatever hope he had of patching up his marriage.

Even in the spring, his misfortunes continued. Patrick had invited some local kids to the grove for a tour of his renewal efforts, hoping to show Big Pine that the grove was down but not out. As he started to give the tour, he realized he needed to move his camper out of the way. He hitched it to his pickup truck and started the engine, but as he pulled the camper forward, he heard a sound that made his blood curdle. It sounded like bones being crushed. With the children's eyes on him, Patrick jumped out of the truck. His dog, Bella, had been sleeping underneath the camper, and he had run her over.

The final sign came a few weeks later when smoke started to drift over Big Pine. A fire that started at a homeless encampment in the island forest surged out of control, feeding on dead trees left from the storm. The local fire department was unable to contain the flames, and by the time reinforcements arrived, it had been smoldering for more than a week.

The fire came within twenty feet of the grove, but it never crossed onto Patrick's property. Even so, it seemed like yet another omen, one more plague for a man who had already lived through death and floods. No matter how hard Patrick worked to restore order, life would never go back to normal. The disaster would never end.

II.

The Last Straw

The thousand-odd islands that make up the Florida Keys are the first flock of canaries in the coal mine of climate change. In the five years since Hurricane Irma, the residents of these islands have been forced to confront a phenomenon that will affect millions of Americans before the end of the century. Their present calamity offers a glimpse of our national future.

Even though the phenomenon underway in the Keys has already altered thousands of lives in the United States, it is so poorly understood and so little discussed that it does not really have a name. Politicians call it "climate migration," and advocates call its victims "climate refugees," but these terms are too narrow to encompass the real scope of the phenomenon. Moreover, none of these phrases can capture the strange and chaotic way that people move in response to climate events, the mixture of voluntary and involuntary relocation that characterizes this growing exodus from vulnerable areas. The process is messier than mere migration, but more profound than mere population turnover; slower than a refugee crisis, but more drastic than a demographic shift.

As many conchs will be proud to tell you, the Keys are one of a kind: there is no other place in the world that boasts the same combination of geological, ecological, and sociological characteristics. The islands have a special, addictive quality about them, an air of freedom that leads people to turn their backs on mainland life. But the Keys are far from the only places in the country that risk succumbing to climate change over the next hundred years, and the dynamics that threaten them are the same ones that will soon threaten much of the nation.

The most obvious of these dynamics is the one that sent millions of Floridians fleeing up the coast in September of 2017. Nature is changing. Today's hurricanes tend to be stronger, wetter, and less predictable than those of the last century. They hold more moisture, speed up more quickly, and stay together longer. It's difficult to tell for certain what role climate change plays in any individual storm, but in the case of Hurricane Irma there is little doubt that the warmth of the Caribbean Sea made the storm more powerful, allowing the vortex to regain strength overnight as it barreled toward the Keys. As global warming continues to ratchet up the temperature of our oceans, we can expect more storms like Irma. The danger to the Keys doesn't end with hurricane season, either: a slow but definite rise in average sea levels over the past decade has contributed to a rise in tidal flooding, leaving some roads and neighborhoods inundated with salt water for months at a time.

On their own, these natural disruptions might not be enough to push a place into a state of collapse. After all, flooding has been a problem for as long as humans have lived near the water, and the chances of a direct hit from a major storm will always be small. But there is another, more powerful dynamic that has thrust the Keys into existential danger, and that is the sheer fragility of human society on the islands. In other words, part of the reason places like Big Pine are collapsing is that they never should have been built in the first place.

For centuries, the Keys were mostly uninhabited, visited only by Indigenous hunter-gatherers or errant explorers, and they might have remained that way had a combination of historical accident and human gall not opened them up to the mainland United States. The first permanent settlement on the islands was not established until 1821, when a businessman named John Simonton purchased Key West for $2,000 at a café in Havana and traveled there to build the nation's next great port city. For much of the nineteenth century, the region's main industry was "wrecking," or the salvage and sale of goods from crashed ships, a trade that made Key West the richest city in Florida. By World War I, though, improvements in ship navigation had led to a drop in the number of shipwrecks, which in turn caused an exodus toward Miami.

In the 1930s, just when the region's economy was at its weakest, a storm much like Irma threatened to erase Simonton's paradise. On Labor Day 1935, one of the largest hurricanes in the history of the United States barreled through the islands, killing as many as six hundred people and destroying an estimated 90 percent of permanent structures. In a strange twist of fate, the storm turned out to be the Keys's saving grace: the commercial railroad connecting the Keys to the mainland had been so badly damaged in the storm that its owners sold the tracks to the state of Florida, which used them to complete a highway down the archipelago. The highway connected Key West to the mainland and hastened its transformation into a vacationer's paradise, opening it up to the fantasies of bards from Tennessee Williams to Jimmy Buffett. Drifters from all over the country filtered down to lower islands like Big Pine to fish for grouper, play in amateur bands, and take quaaludes.

As Buffett's "Margaritaville" became a jukebox staple and a generation of retirees cashed out their pensions, thousands of people around the country decided they wanted a slice of paradise. This tide of snowbirds pushed real estate development out from Key West and into the smaller islands that make up the midsection of the archipelago. To satisfy demand for new homes, developers employed building techniques that were as shortsighted as they were audacious: some sliced through seaside mangrove forests and shoved fake land out into the water, while others dredged canal networks where vacationers could park their boats. This tide of development reshaped not only larger and more populous islands, but also smaller and lower-lying spits, places where humankind had never so much as pitched a tent.

Over time the rum-soaked anarchy of the undeveloped Keys got harder to find, until finally it became isolated on and around Big Pine, one of the few major islands that managed to resist a full developer takeover. To this day, most of Big Pine is an untouched conifer forest, and many residential roads are still unpaved. This wilderness has helped suppress property values, keeping the island affordable for the line cooks, fishermen, and boat buffers who have been priced out of Key West by the tourist boom they help sustain. It has also attracted more than a few hippies, ex-hippies, and outlaws—a frequent joke is that everyone there is either running from something or toward something.

The possessive love that many conchs have for the Keys can make it hard to remember just how short and tenuous the history of the islands has been. The first significant development on Big Pine did not arrive until seventy-five years ago. Almost nobody who lives there was born in the Keys or even in Florida, and most current residents are thirty or forty years older than the neighborhoods they inhabit. Furthermore, the freebooting society that exists in these places has been threatened almost from the moment they came into being. Decades of development in the Keys have pushed rents up to New York City levels, and the much-maligned cruise ship boom has blunted the island's Wild West spirit. If the island has always attracted people living on borrowed time, the island itself has been living on borrowed time as well.

In the five years since Irma, the bill has come due, not just on Big Pine but for the entirety of the Keys. The hurricane made undeniable what previous floods had only suggested: that climate change will someday make life in the archipelago impossible to sustain. The storm was the first episode in a long and turbulent process of collapse, one that will expand over time to include market contraction, government disinvestment, and eventually a wholesale retreat toward the mainland. Irma may not have destroyed the Keys in one stroke, but the storm ran down the clock on life on the islands, pushing their residents into a future that once seemed remote. The impulse to stay, which once bespoke a conch's devotion to his or her adopted home, now looks a little more like denial. The decision to leave, on the other hand, which once signified surrender, now looks more like acceptance of the inevitable.

—

The aftermath of a major disaster looks a lot like a gold rush. Thousands of people arrive at the disaster site from all over the country representing every branch of government as well as innumerable nonprofits, charity groups, churches, private companies, and contractors. After that come the evacuees, the thousands of people who left home during the hurricane or wildfire or tornado and are now trying to return to inspect the damage. Everyone seems to be selling something, or offering something, or trying to help someone get in touch with someone else. The rush of people is so large and so chaotic that most people have a hard time figuring out who, if anyone, has the power to make things better.

The most important player, though, is the federal government, which shoulders most of the burden for responding to large disasters. At least in theory, the way it works is that the government shows up after a storm and hands out money: FEMA distributes individual grants to people in affected areas, the Small Business Administration (SBA) offers low-interest loans to help victims rebuild, and the National Flood Insurance Program (NFIP) pays out claims to people who have flood insurance. The feds also reimburse state and local governments for the cost of cleaning up debris and infrastructure. All this money acts to rewind time and restore everything to normal.

Things almost never work out that way, but in the Keys it was not even close. FEMA was already shorthanded after having spent the previous few weeks responding to Hurricane Harvey in Houston, and the remote nature of the Keys made it difficult for the government to bring in the requisite aid as fast as was necessary. Meanwhile, the members of Big Pine's off-the-grid population found it almost impossible to jump through all the bureaucratic hoops required to receive federal assistance: many of them did not have a permanent residence, or did not have a deed to their home, or had never known they needed to buy flood insurance, which meant that FEMA, the SBA, and the NFIP denied their relief applications. The result was that even months after the storm, most of the island still looked like Grimal Grove: torn apart, bereft of life, ramparted on all sides by piles of stinking debris.

In the absence of government assistance, the hard work of rebuilding the island fell to a small group of hardened Piners, locals who refused to let their island remain in a state of disrepair. One of these people was Lenore Baker, a bartender at a local seafood grill who started a volunteer cleanup effort in the months after the storm. Lenore was fifty and single, and like many of her neighbors she considered leaving the Keys for good when she saw the piles of trash that lined every road in Big Pine. When at last she decided to stay, she figured she'd better take the cleanup into her own hands: she called up a few of her friends and posted on Facebook that she was going down to the main highway the next day to pick up trash. More than two dozen people showed up to help, lifting plywood boards and boat masts out of the weeds in eighty-degree heat. It wasn't just hardened locals, either, but snowbirds who had flown down to fix their vacation homes and seniors who wanted to lend a hand. They picked up fridges full of rancid fishing chum, split cabinets with crushed china dishes inside them, and hundreds of pounds of brown sea gunk. Lenore held another cleanup the next weekend, and another one the weekend after that, and another one the weekend after that. As she and her crew worked their way along the length of US 1, they drew hundreds more volunteers.

Recovery efforts have since acquired an almost mythical quality on Big Pine. Many island residents had already been skeptical of government even before the storm, but FEMA's shortcomings only further convinced them of the incompetence of the federal bureaucracy. As they tell it, individual heroes like Lenore are the only reason the island ever recovered, the only reason why there aren't still piles of debris lining the side of every road.

If Lenore succeeded in restoring the body of the island, though, its spirit was still languishing. For every local who joined one of her cleanups, there were ten more like Patrick Garvey who were struggling through a dark night of the soul. As autumn gave way to winter and winter surged into a frothing spring, more and more people on Big Pine felt their grip on the island start to slip away.

Witness to this darkness was a bartender named Debby Zutant, who worked at a smoke-filled dive bar called Coconuts. Born in Syracuse, Debby is another classic Big Pine figure—she describes herself as an "extreme empath" and was once married to her half brother, whom she met in her thirties after growing up with adopted parents. Debby was among the first people back to work after the storm, and Coconuts was the first business on the island to reopen.

In the months after the storm, the men of Big Pine started to spend long hours in Coconuts, entering the bar in the middle of the afternoon and staying put until late in the evening. Once the aid organizations and church groups flew home, there was nothing left to prop up the community. There was nowhere to work, nothing to see, nothing to do but sit around and wait for aid money to arrive. Debby spent hundreds of hours in what she called "group therapy" with the Coconuts crowd during that first winter, pouring up free drink after free drink and trying to help them work through problems that seemed to get worse by the day.

The remaining Piners found themselves living in a place that was nothing like the one they remembered. The renegade spirit of the island had all but vanished: gone were the anarchic trailer park communes, the impromptu confabs in the Winn-Dixie parking lot, the earthshaking parades of the local motorcycle club. The church congregations broke up, the fishing groups disbanded. The spirit of toil and solidarity that took hold of the island in the storm's wake soon vanished, replaced by a dread about the island's future. Debby tried her best to keep tabs on all her customers, often calling to check on them after they left, but she could only do so much to ward off the darkness. When her customers walked out of the bar and back to their cars, the wreckage of their lives was always waiting for them.

One of Debby's closest friends was a Coconuts regular who ran a home-grown taxi company on the island; he was what she calls "an old-school Boston gangster," not the kind of person who is easily rattled by adversity. Before the storm he had lived in the Seahorse RV Park, a small motor home community that in many ways was the spiritual heart of Big Pine. "It was a den of iniquity, too, but we won't talk about that," says Debby. Like other Seahorse residents, Debby's friend had evacuated during the storm, but when he came back, he learned that the county government had condemned the trailer park, deeming it too dangerous to rebuild. With nowhere to live, he gave up his trailer and rented a room elsewhere on Big Pine. Much of the island's population was gone, and his taxi business slowed to a halt.

One night in February, when Debby's friend closed out his tab at Coconuts, he left Debby a fifty-dollar tip. She came around the bar to give him a long hug.

"You're one of my favorite people in the world," she told him. At the time she didn't think anything of the tip—that was just what people did for each other.

The next morning, Debby got a call. Her friend had shot himself.

As the months wore on, the dread that had descended over the island claimed still more victims. At least a dozen people committed suicide in the winter and spring that followed the storm, many of them Debby's patrons at Coconuts. By the time summer came around and the tourist season resumed, it felt as though the Rapture had taken place on Big Pine, leaving behind only those who were desperate or crazy enough to believe they could rebuild what had once existed there. The official statistics say that just under 5 percent of Keys residents left the archipelago for good after Hurricane Irma, but between the deaths and displacements the number of absences on Big Pine felt much larger. There were still people like Patrick and Lenore straining to rebuild their lives, but the space between them seemed to have grown larger. Some days, as Debby drove around the island, it felt like she was the only one left.

III.

Never Going Back Again

T he term "climate migration" is an attempt to explain why people leave one place in favor of another; it assigns a primary motivation to movements that may be voluntary or involuntary, temporary or permanent. Yet even if the primary cause for an episode of migration is clear, there are still countless other factors that influence when, where, and how someone moves in response to a disaster. It's this messiness that is reflected in the word "displacement": the migratory shifts caused by climate change are as chaotic as the weather events that cause them. The migration that took place in the Keys after Irma was just such a chaotic displacement event: some people left by choice, but many others left because they had no other option.

For some families the decision to depart the Keys was easy. The storm was a traumatic event, more than enough to convince many people that life on the islands was too dangerous to accept. They came back home, fixed up their houses, and got out. That was the case for Connie and Glenn Faast, who left the island city of Marathon for the mountains of North Carolina after spending almost fifty years in the Keys.

"It was pretty much immediate," Connie told me. "It's just too hard to start over when you get older. We couldn't risk it."

The Faasts had lived the kind of life you can only live in the Keys:

Connie worked on commercial fishing boats and in a local aquarium, while Glenn owned a boat maintenance company and raced Jet Skis in his spare time. They had stuck it out in the Keys through several major storms, including 2005's Hurricane Wilma, which brought five feet of water to their little island and totaled three of their cars; Connie still shudders when she remembers the image of her husband wading through the water around their house with snakes climbing all over him, clinging to him for shelter from the flood. The Faasts had second thoughts after that storm, but the Keys were paradise, and besides, they didn't know where else they would go.

When Irma came twelve years later, though, the choice was much easier. During the evacuation, it took the Faasts a week to find a decaying hotel in Orlando where they could wait out the storm. As the hurricane passed over the center of the state, it knocked out their power, leaving them and their pets to spend the night in one-hundred-degree heat without air-conditioning. "That was it for us," she said. They had to get out—not just out of the Keys, but out of Florida altogether.

When they returned to Marathon, they discovered that their home was the only one in the neighborhood with an intact roof; even a mansion next door to their house, built only the year before, had had its top rolled back like a sardine can. They put the house on the market as soon as they could, but it took a year for the place to sell, in part because property values had risen so steeply that most people in the area couldn't afford to buy anything. The storm had scared many people off, but it had also destroyed a quarter of the Keys's housing stock, which drove up prices for the homes that survived. In the meantime, the Faasts saw their friends start to leave as well: one moved to Sarasota, another to Orlando, and a third friend, who had been the first-ever mayor of Marathon, talked about moving to central Florida.

"We thought it would be devastating when we left," Connie said, "because we love the Keys. But when we pulled out of there, we were so, so relieved."

Hundreds of people like the Faasts left the Keys of their own volition in the years after Irma, deciding one way or another that the risks of staying there outweighed the benefits. But the other and more turbulent phenomenon after the storm was the involuntary displacement caused by the shortage of affordable housing on the islands. The storm destroyed not only the massive mobile home parks on islands like Big Pine, but also hundreds of so-called downstairs enclosures, small apartment-style units that sat beneath elevated homes. It also wiped out dozens if not hundreds of liveaboard boats and older apartment complexes in island cities like Marathon. These trailer parks and apartment complexes had been havens for resort waiters, boat buffers, and bartenders, allowing them to get a foothold in an archipelago that had long ago become unaffordable for anyone who wasn't rich. Now all that housing was gone, and FEMA's 50 percent rule prohibited most trailers and downstairs enclosures from being rebuilt. Many of those who had been lucky enough to own small homes or campers hadn't been able to afford insurance, which meant they missed out on the payouts that went to wealthy homeowners and part-time vacationers. To make matters worse, the government of the Keys couldn't build enough new homes to fill the gap created by the storm: the state had long ago imposed a de facto cap on the number of building permits the county could issue, an attempt to make sure the population did not grow too large to evacuate the islands in a single day. Thus it was impossible for most conchs either to rebuild their old homes or buy new ones. Some of those who lost their homes were able to crash with friends and family, and others got by living in tents or trailers, but others resorted to the forest homeless encampment where the Big Pine wildfire started. The lack of housing made the storm survivors feel as though they were stuck in a permanent limbo: life on the islands became a game of musical chairs, in which only the highest bidders could end up with a seat.

Debra Maconaughey, the rector at St. Columba Episcopal Church in Marathon, spent the years after Irma trying to forestall this involuntary displacement. When the storm hit, Maconaughey and much of her congregation were in Ireland, retracing the steps of the original St. Columba, and by the time they returned to the Keys it was clear that housing would be the defining challenge of the next few years.

"Everybody's house was destroyed," she told me. "That's what people would need the most." We were speaking in the church's open-air pavilion, where Maconaughey had been delivering outdoor sermons even before the coronavirus pandemic. Irma had weakened the timbers that supported the roof of the chapel, forcing the church to move outside.

In the first week Maconaughey was back, she helped transform St. Columba's campus into a massive shelter for boaters who had lost their homes in the storm, cramming two dozen air mattresses into a loft that had previously been used for an after-school program. No sooner did the displaced boaters arrive at the campus than they started to contribute to the relief efforts themselves, carrying debris out of nearby houses and cutting through fallen trees with chain saws. The next week, Maconaughey and her congregation installed approximately two dozen trailers around Marathon, giving the boaters a long-term place to stay.

Maconaughey knew there was no chance the county government would restore all the housing that had been lost in the storm, but after a year went by, she found herself shocked at how little had been rebuilt. A nonprofit land trust had erected only a handful of new cottages and a $50 million state program called Rebuild Florida had repaired only two homes, a pittance compared to the thousands of dwellings that had been swept away. So Maconaughey called up the nonprofits that were funding St. Columba's relief efforts and made an unconventional proposal: the church, she proposed, would buy some derelict housing and fix it up. She had her eyes set on a leaky, mold-filled apartment complex in Marathon that had been condemned for sewage issues a few years earlier. The apartment complex finally opened in the summer of 2020, providing cut-rate housing to sixteen families who had been staying on couches or in trailers since the day the storm hit.

But for every person who found permanent shelter, there were more who could not afford to wait for the islands to recover. This wasn't only because people didn't want to return, but also because there were no homes to which they could return. Maconaughey told me with distaste that in several places along Marathon's beachfront, developers have built single large mansions on lots that once contained three or four small homes each. The lack of affordable housing in turn created a labor shortage: fire and police departments couldn't find enough officers to fill their shifts, boat maintenance companies struggled to locate buffers and repairmen, and many hotels went shorthanded through the on-season rush. When employers exhausted their hiring options on the islands, Maconaughey said, they started to hire workers from the mainland towns of Homestead and Florida City, who take a two-hour bus ride in either direction to work for minimum wage.

"I think people are really struggling, and it's just below the surface," she said. "We're a tourist area, so it's in our best interests to make it look nice from the highway, but there's hidden pain." During the early months of the recovery, a representative from a church relief group told her it takes three to five years to recover from a disaster. At the time that sounded preposterous to her—the Keys would be back on their feet within a year, she thought. But even as the five-year anniversary of Irma approached, she wasn't sure the old Keys would ever come back.

Maconaughey told me about the church sexton, Mike, who was driven out of the Keys by Irma. Mike showed up after the recession in a homeless shelter in Marathon. He was blind, and when he first arrived at the shelter he couldn't take a shower or put on clothes without assistance. After a year in the shelter, Mike started attending services at St. Columba, and soon displayed a great talent for weaving wooden canes and chairs, a craft he often practiced on the church pavilion after sermons. He also taught the kids in the after-school program how to play chess.

Mike was on the Keys as the storm approached, not with the congregation in Ireland. He first sought refuge in the massive Miami hurricane shelter, but by the time he got there, that shelter was full. As shelters in Florida all reached capacity, emergency officials herded evacuees from the Keys up toward Georgia, North Carolina, and Virginia, offering them bus transportation as far as they were willing to go. Mike was unsure when he would be able to return to the Keys, so he asked for a ticket to Minnesota, where he grew up. He was never able to get back.

"We kind of lost him," Maconaughey said. "He got on a bus to evacuate and now he's gone. He was a huge part of our community. . . . You have to ask yourself, do you ever recover from something like this?"

IV.

Time and the Tide

The existential threat that faces the Keys is easier to understand when you remember that from the standpoint of geological time the islands are merely a blip. Just as communities like Marathon and Big Pine have been threatened since the moment they were founded, the islands themselves have been sinking for as long as they have been above the water. Even in the absence of another major storm like Irma, the threat of rapid sea-level rise throws their very survival into jeopardy.

The Keys formed as an underwater coral reef just 130,000 years ago, at a time when ocean levels were twenty-five feet higher than they are today and when much of Florida was submerged. During the last ice age, water from the Atlantic froze into a glacier that covered much of North America, lowering global sea levels and leaving the reef exposed. Forests flourished on what had once been the ocean floor and the reef structures fossilized into mountains of limestone, which later became covered with soil and minerals. Around fifteen thousand years ago, the ice age ended and the glacier started to melt, dumping more water into the oceans. The ocean started to rise again around the reef, leaving only a long chain of limestone islands poking above the water. The Keys as we know them have only existed during a brief and improbable equilibrium between massive ocean disturbances.

That equilibrium is now coming to an end. Global warming has already locked in as much as six feet of sea-level rise in the next century, a surge that will disfigure coastlines all over the world and force millions of people to move back from the shore. In the Keys, the prognosis is even worse. Many of the islands in the archipelago, perhaps all of them, could go underwater altogether by the end of this century. More so than almost any other place in the United States, they are doomed.

The enormous task of recovering from Hurricane Irma has now been supplanted by an even bigger task, that of preparing the islands for an era of constant flooding from high water. There will be no rush of FEMA aid for this crisis, though, because federal law does not recognize the slow rise of the oceans as an emergency like a hurricane or a wildfire. In the absence of a coordinated federal response, county and state officials have proposed a wide variety of solutions for sea-level rise. Some have proposed hunkering down and ignoring the problem, others have proposed transforming the islands with flood walls and barriers, and others have proposed paying residents to move somewhere safer. Faced with the magnitude of the crisis, though, the local government of the Keys has had no choice but to tackle the impacts of climate change on a first-come-first-served basis, restricting its attention only to the problems that are already staring it in the face. Right now, the problem is roads.

In November 2020, county officials presented the results of a years-long study that attempted to calculate the future impact of sea-level rise on the roads in rural and unincorporated parts of the Keys. The residents who lived on such roads already knew that sea-level rise was a problem—low-lying streets often went underwater for days at a time, sometimes forcing residents to traverse their blocks by kayak—but the county had never made a formal attempt to gauge just how bad the issue might get.

Leading the effort was Rhonda Haag, the chief resilience officer for Monroe County, which encompasses the Keys. Haag is mild-mannered, with a friendly midwestern accent, and even when discussing the grim future ahead of the islands she couched the bad news in terms of pragmatism—study, deliberation, decision. She had made national news a year ealier when she attended a climate resilience conference and offered a brief peek at the roads study, focusing on a rural stretch called Sugarloaf Key. Haag estimated that it would cost about $75 million, or $25 million per mile, to protect the road from sea-level rise through 2025—an impossibly steep price considering that there are fewer than a dozen houses on the road.

The findings of the full study were even more dramatic. Haag's office rated every roadway in the Keys based on its vulnerability (exposure to flooding) and criticality (importance to residents and commerce), producing a ballpark estimate for how much money it would cost to protect all county-owned roads from flooding. The study, which excluded cities like Marathon and Key West and focused only on unincorporated land, found that almost half of all roads were vulnerable to sea-level rise, and that protecting them would cost a jaw-dropping $1.8 billion, far more money than the county could hope to raise on its own. Even that sum would only protect the roads through the year 2045, at which point the county would have to find even more money to fortify them all over again.

As part of the study, Haag's team created a series of color-coded maps showing which areas of the Keys would suffer the most with successive stages of sea-level rise. The maps revealed that by mid-century almost no corner of the archipelago would remain safe: even two feet of sea-level rise would push the water onto the roads in Lenore Baker's neighborhood of

Big Pine; another foot would push tidal flooding onto the roads around Grimal Grove. Even if residents elevated their homes, life would be almost impossible if the roads were not elevated.

The enormity of future sea-level rise threatens to force policy makers like Haag to decide which areas are worth saving and which are not. A truism in the field of climate adaptation is that you can save money later if you spend money now: one dollar of adaptation spending in Miami, for instance, might save six dollars in flood damages for the city down the road. A recent study, however, found that that logic did not apply in the Keys: one dollar of adaptation spending on the islands yielded only forty-one cents of future savings. In sheer numerical terms, the islands don't have the population to justify the money that would be needed to protect them.

"When you look at that price tag, you have to start determining which roads you're going to elevate," Haag told me when I spoke with her on the phone. "Are you going to allow some level of water? If you don't elevate them as high, you can spread your dollar further, but that doesn't last as long."

After completing the study, Haag's next task was even harder: over the following year she had to dialogue with the public to create a list of the highest-priority roads, then try to figure out which of them the county had the money to elevate. Recent federal infrastructure bills promised to provide some money for road repair, but Haag knew federal grant money wouldn't be enough on its own, so she thought about proposing a special sales tax for raising roads, or annual property taxes on residents in vulnerable areas. What would happen to people who live on roads that didn't make the repair list was an open question. Haag thought perhaps the county could run seasonal ferry services in the worst-affected neighborhoods, helping connect residents to the rest of the Keys during the worst months of flooding.

In some parts of the Keys, this future may have already arrived. Every autumn the Eastern Seaboard sees a series of extra-high tides known as "king tides," which occur when the moon is at a near point in its rotation around the earth. In previous decades these tides have never been more than a nuisance, but over the past five years they have inundated a pair of Key Largo neighborhoods that were built on artificial land. In 2019, the communities of Twin Lakes and Stillwright Point, about a hundred miles east of Big Pine, were both flooded for more than ninety consecutive days with low-lying water that sloshed in from Florida Bay. Most houses in the communities are already elevated, but if residents didn't have pickup trucks with high suspension, it wasn't safe for them to drive out to the main highway.

When I drove through Stillwright Point a year later, the county had placed flood barriers—they looked like long orange burritos—to protect the streets from flooding, but the tides had crept in anyhow, covering several laneways with three or four inches of khaki-colored water. According to Haag's map, another foot of sea-level rise would mean almost year-round inundation. One resident from Stillwright Point told me that residents in his neighborhood have already begun to jockey with residents from Twin Lakes over which community will be the first to get its roads raised. The county expects to get additional funding from federal and state infrastructure measures passed during the pandemic, but the expectation in Haag's office is that there will not be enough money to save every place. Sooner rather than later many people in the Keys will need to give their homes back to nature.

Haag admits that she's advocating something that has never been done before, something she's not even sure she *can* do: she has assigned her staffers to research legal precedent for intentional road abandonments, anticipating a potential lawsuit. To the extent that precedent exists, it is scant, and if the abandonments really do happen, then the county could face blowback from irate homeowners. The alternative to this plan, though, is to do nothing, and let residents make their own sporadic exodus from vulnerable areas.

Haag is an optimist, a necessary qualification for someone with "resilience" in her job title. She believes that if sea-level rise stays below a certain threshold—that is, if the islands do not go underwater altogether—then the Keys can be saved for human habitation. Even so, the place that remains won't look anything like the place that most residents know today.

"It could be a matter of alternative living," she said. "If they don't mind water under their homes, maybe they can find another way to get to their house with a boat. What we're trying to say now is maybe we can learn to live with water."

—

Living with water is easier said than done, especially for those who have seen what water can do.

On my last day in the Keys, I paid a visit to Patrick Garvey. When I entered the grove through the back gate, I found him dressed in a T-shirt and gym shorts, hauling a two-by-four across the yard. He was working on rebuilding the grove's outdoor kitchen, hoping to open the space up for events. A few feet away, some teenagers were pulling weeds in an elongated flower bed. When Patrick finished work for the day, he and I sat down at a picnic table outside his camper and watched the sun set.

Like the rest of the Keys, the grove was caught in a limbo. Patrick thought every day about the possibility of another storm, and those thoughts often made his repair work feel futile—"Irma's still in my head," he told me. Even if the Keys never saw another storm, the grove might not always be viable: the rising tide was pushing brackish water farther into Big Pine's soil, blanching several acres of forest around the island's northern edge. When the water made its way beneath the grove, Patrick's breadfruit colony would die.

For the moment, the grove was alive, but Patrick's dreams for it were more fragile. He had no family and few friends on the island, and most days only his new dog, Luna, kept him company. The twins had stayed with him for the first part of the pandemic, during which time he'd built them a tree house, a zip line, and a swimming pool, but they went back to Brazil at the end of the summer. Living in the Keys without his daughters made Patrick start to question what he was doing there. When he bought the grove, he had imagined it as a place where he could put down roots, so to speak. The storm made that impossible.

"I came here to build a family," he told me. "If I can't get my kids here, there's no point in doing any of this. It's all meaningless."

As darkness fell around the grove, Patrick's banged-up camper became the sole source of light on the property; mosquitoes started to land on our arms. Patrick nudged Luna awake at his feet and stood up, shifting toward the camper. He had to change clothes and drive down to Key West to check on a property he was house-sitting—he was still doing odd jobs to supplement the income he got from the grove.

Before we parted ways, Patrick told me he was thinking of selling the grove. In an ideal world, he said, he would find a conservationist or horticulture enthusiast who could take over the breadfruit forest, but he would be just as satisfied if the property became a vacation rental for tourists. He was still afraid the property would be on his hands when the next hurricane comes, and he was less certain than ever about staying in the Keys.

Before Patrick could sell the grove, though, he had one last hurdle to clear. Decades ago, when Adolf Grimal first got to Big Pine Key, there was a road running right through what is now the middle of the grove. Grimal ripped up the road and extended the grove over it, but the county still claimed ownership to a tiny piece of Patrick's property. For Patrick to put the grove on the market, he had to convince the county to release the road from the public trust.

The county granted the abandonment the following summer, leaving Patrick free to sell the grove and leave the Keys. In one sense this abandonment was just a matter of paperwork. Patrick filed a petition, a county clerk approved it, and the road got erased from the record books. In another sense, it was a dress rehearsal for the retreat that will become necessary across the Keys and elsewhere in the United States. The county let go of the road, Patrick let go of the grove, and Big Pine let go of Patrick. What happened next was for the water to decide.

CHAPTER TWO

After the Flood

MANAGED RETREAT

AND ITS VICTIMS

Kinston, North Carolina

Ghost Town

L incoln Street is quiet. There are no cars on the road, and, despite the nice weather, every front porch is empty. Half the lots on the street are vacant, and more than half the houses that remain standing have long since been condemned. Some of these abandoned homes have plywood boards over their windows, while others have their house numbers spray-painted across the front facade. Still others have gaping holes where their doors used to be, allowing passersby to see straight through to the dense forest beyond. Mottled beech trees have surged up over home foundations, tangles of dead wood stretch over what remains of the asphalt, and dead leaves gush up through the storm drains like geysers. Farther on, a dozen more fossilized streets extend to the left and the right. Here and there is a relic of human habitation, a punctured tire or maybe a mold-stained mattress, but for the most part nature reigns unchallenged. A pack of wild orange dogs roams one end of the abandoned neighborhood, growling at anything that gets too close. Elsewhere, the streets are weighed down by a pulsing, ponderous silence.

This labyrinth is all that remains of Lincoln City, a historic Black neighborhood in the city of Kinston, deep in the swamps of eastern North Carolina. The first thing that anyone who lived in Lincoln City will tell you is that the neighborhood felt like one big family. Everybody knew everybody. Nobody locked their doors. If you went hungry in Lincoln City, it was by choice; on the other hand, if you acted out at a friend's house, your mother would know about it before you even got home. Lincoln City was not a wealthy place, but its residents never wanted for anything: most of the area's eight hundred families held manufacturing jobs, or worked in the local tobacco prizeries, but there were also doctors and lawyers in the neighborhood, as well as dozens of teachers and social workers. The neighborhood had its own elementary school, its own community college, an auto repair shop, and two stone yards. There were at least a dozen grocery stores of varying sizes, some selling fresh produce and others hawking candy and cookies to neighborhood kids. Many residents grew herbs and vegetables in their backyards, picking and roasting them on Sundays. The homes were mostly shotgun houses, brick cottages if you were lucky, but for the families who owned them these houses were a source of tremendous pride. That the grandchildren of enslaved people and sharecroppers could own their own property was considered an almost unaccountable blessing.

What people remember most about Lincoln City is childhood. Every afternoon, when school let out for the day, dozens of kids tore down Lincoln Street on bikes or bare feet, winding in and out of backyards and around the street's three big churches. They shot basketballs on the local court, played softball in a field with makeshift bats, or spent lazy weekend nights burning tires. The long walk back from the local high school provided teenagers with ample opportunities to court each other, and if they needed privacy they could always sneak off to the "sand hole," an expanse of beach-like sand down by the banks of the Neuse River. In the summer you could get candy from one of the convenience stores, taking it on credit if your parents didn't have spare cash; failing that, you could hoof it to a nearby housing project, where mothers made a freeze-ice concoction known as "hard cups," which came in any flavor you wanted. For the children who lived there, Lincoln City was a world unto itself. It was hemmed in on the north by an expanse of housing projects, on the west by a white-dominated downtown, and on the south and east by the sluggish turns of the Neuse River. There was nowhere else to go, but where else would you want to go?

That was a long time ago, and Lincoln Street has now been quiet for more than twenty years. In 1999, a massive hurricane called Floyd caused dramatic flooding throughout eastern North Carolina, swelling the Neuse River and drowning Lincoln City beneath almost five feet of water. A few weeks later, FEMA descended upon Kinston and made city leaders an offer they couldn't refuse: the federal government bought out the riverside neighborhood, giving each homeowner a check for the value of their home so long as the owner agreed to vacate the floodplain and move somewhere else. Once the residents of the neighborhood were gone, the government knocked down all the homes, preventing future flood damage and saving itself money down the road. By the time the next storm came around, Lincoln City was gone—or almost gone. The overgrown labyrinth around Lincoln Street still serves as a monument to the federal government's first experiments in coordinated climate migration, a reminder of what we stand to lose when we sacrifice some communities for the safety of others.

The story of climate displacement is messy, diffuse, and many-stranded. There is no single starting point, no shot-heard-round-the-world moment that kicked off this evolving process of relocation. Any beginning must be arbitrary, and Lincoln City is as good a place as any other to begin. The story of the neighborhood is the story of how, after centuries of unmitigated growth, the federal government began to halt the cycle of destruction and reconstruction in flood-prone areas and move people away from the riskiest places.

It is the story of how we began to retreat.

II.

The River Gives,
the River Takes

E ven before the floods, the first residents of Lincoln City were already
refugees of a kind. In the summer of 1865, after the Civil War ended,
thousands of newly emancipated slaves walked off their plantations and
into a remade world. Most of these free people stayed within a few dozen
miles of the plantations where they had once been in bondage: few had the
resources to travel all the way to the industrial cities of the North, and the
biggest employers in their native states were plantation owners, who often
rehired their old slaves to work the same cotton fields. The free people
who chose to stay in the areas where they were born had to learn to survive
within a local economy that had been almost entirely dependent on their
unpaid labor.

A few decades after emancipation, a Black preacher's son named Lincoln Barnett bought a wide swath of land outside the city of Kinston. The
white entrepreneur who previously owned the land had sought to build a
planned community there, but had found that his neighbors were unwilling
to take up residence in what was essentially a swamp. In Barnett's hands,
though, the tract soon became a haven for dozens of freedmen who had
migrated to the city to work on the docks.

From the beginning, the fate of the community was intertwined with
the river on whose banks it had been founded. William Lawson, who was
born in the neighborhood in the 1920s, told me that the first residents
settled there because the moist soil was good for growing crops. Lawson's
parents had both been tenant farmers, one generation removed from slavery

themselves, and they were among the first couples to settle on the fertile land that later became Lincoln City. The couple built a home on the closest street to the river, University Street, which back then was little more than a dirt road. They grew enough food in the backyard to feed William and his six siblings.

"Back then, there were only two houses on University Street," Lawson told me. "If you saw anyone walking beyond those two houses, you knew that they were on their own journey." There was a single shop on the street as well, a "backyard store" that sold soap and other general goods, but for almost half a mile beyond that there was nothing but dark, rich soil. The early residents of Lincoln City and the other "bottom" neighborhoods learned to tolerate the occasional incursion of water after a heavy rainfall. Much later, when the neighborhood had been torn down, a resident born in 1924 was quoted as saying that "all my life I was accustomed to avoiding the floods, moving out if I had to, and go[ing] to higher ground"; Lawson recalled occasional flash floods as well. These were benign events, though, especially because there was so much open soil to soak up all the water. At worst, the river water would stop just short of most people's doorsteps and drain out after a few days.

Despite the bucolic privacy of the bottomlands, the generations of Lincoln City lived in the shadow of white supremacy and systemic racism, and there were few if any jobs in the area for Black people. In the early part of the twentieth century, Kinston was a hub for North Carolina's tobacco industry, but Black people were unwelcome as employees in the local prizeries, and the biggest driver of "tourism" near Lincoln City was Sugar Hill, a red-light district that attracted clients from around the state. Lawson wanted to change that. After he attended college in New York City, he moved back to Lincoln City and took a position with the state department of economic development. He and his colleagues at the state government strove to anchor local industries in Black neighborhoods like Lincoln City, but their efforts bore scant fruit. In a cruel twist, the neighborhood's economic salvation came from a company with a long legacy of contributing to environmental injustice: the chemical conglomerate DuPont opened a polyester factory just a few miles from Kinston in 1953, hiring many workers from Lincoln City. The company has paid out millions to settle lawsuits alleging it dumped toxic chemicals such as PFAS into major waterways like the Ohio River.

The federal government also neglected the needs of the river community. The federal agency responsible for flood prevention was the US Army Corps of Engineers, a quasi-military public works agency that over the course of the twentieth century had adopted a lead role in taming the nation's waterways. Taking orders from Congress, its engineers dredged channels and built levees from Mississippi to New Jersey in an effort to control flooding and protect bastions of tourism and agriculture. In the late 1960s, the Corps planned to build a pair of dams along the Neuse River. The first dam, on the upstream section of the river, would protect the burgeoning city of Raleigh; the second dam, forty miles farther downstream, would protect Kinston and other cities. The Corps secured funding for the first dam, and it finished the structure in 1981, but never ended up building the second dam, near Kinston. The city was small and stagnant, and unlike in Raleigh, the most vulnerable land there was also the least valuable: thanks to decades of discriminatory housing practices, the flood-prone territory near the Neuse was also home to most of Kinston's Black population. Even if city officials in Kinston had wanted to pursue flood protection without federal support, they didn't have the money for projects like flood walls, and the Black residents of Lincoln City lacked the political clout to gain the Army Corps's attention on their own. The second dam project fell by the wayside, as did other proposals to create new reservoirs in the area.

The inaction lasted through the end of the century, with one exception. When a series of floods in the early 1980s caused heavy damage to a red-light district of Kinston called Happersville, the city took advantage of the disaster to engage in an impromptu urban renewal project. Happersville had for decades served as a kind of open-air waterfront bordello, drawing gamblers and criminals from all over the region, and now the city wanted to ensure it never recovered from the flood. Officials dipped into the town budget and came up with half a million dollars to pay off the impoverished residents of the district, giving them a cash stipend to move somewhere else. It was a slapdash effort, more glorified eviction than climate adaptation, but it worked: Happersville emptied out, the city abandoned the land, and within a few years the neighborhood had sunk into the folds of the Neuse. The buyout was the city's first real attempt to pull back from the river, and it foreshadowed a pattern that would soon emerge all over the country: the burden of relocation would fall on those who had the least ability to resist it.

—

In September of 1996, a midsize hurricane named Fran caused major flooding in Kinston and in surrounding Lenoir County. The damage in Lincoln City forced half the neighborhood to evacuate for weeks as dark brown water crept up to their doorsteps.

The emergency management director for Lenoir County at the time was Roger Dail, a lanky and mustachioed man who wouldn't be out of place in a Coen brothers movie. The day after the storm, Dail got a phone call from an urban planner who worked for Kinston. The planner told him about a new federal program that offered grants for home buyouts in flooded areas. The county could submit a funding application to the state, the state would pass it up to FEMA, and FEMA would pick up 75 percent of the tab for the project. The program had been used a few years earlier to buy out homes along the Mississippi River, and it seemed to be working well.

Dail liked the sound of that. Lincoln City was just as vulnerable to flooding as the red-light district of Happersville had been, but neither the city nor the county had enough money to buy out Lincoln City the way the city had done for Happersville. The residents of Happersville had been a few dozen drifters living in shacks and shanties; Lincoln City, by contrast, had hundreds of families and dozens of stores. The new federal program made the buyout math a lot easier.

Like many people who have influenced the course of history, Dail was just doing his job. It was his responsibility to reduce future flood risk in the county, and it seemed like the FEMA program was the cheapest and most effective way to do that. He got started on the paperwork, and so did his counterparts at city hall in Kinston; they targeted a few hundred homes in Lincoln City and other flood-prone neighborhoods along the Neuse.

The new buyout program cut against decades of disaster policy, which had always underwritten the risk of construction in vulnerable areas. For most of the preceding century, developers had been moving *into* floodplains, not out of them. Real estate tycoons in coastal states like New Jersey and Florida had erected thousands of houses on beachfronts and barrier islands, and engineers in inland areas had drained countless rivers and swamps to allow for construction on formerly uninhabitable land. The rapid pace of this construction was in part the result of ignorance about how floods worked, but it was also the result of an economic dynamic that sociologists call the "growth machine": more construction meant more money for developers, more people living in waterfront towns and cities, and more tax revenue for local governments. A booming tax base allowed more spending on public services, which in turn attracted more people, created more demand for construction, and spread more money around. At least in the short term, everyone in a community won.

The federal government served as the de facto protector of this growth machine, a task it fulfilled through two major agencies. The first was the US Army Corps of Engineers, the nation's principal builder of flood walls, levees, artificial beaches, and other structures that protected people from water. The other major player was FEMA, a younger agency that represented a mishmash of several different departments and distributed material aid after disasters. Both these agencies had always endeavored to keep flood prone communities where they were, not move them elsewhere: the

Corps wrapped levees and seawalls around existing towns to protect them from flooding, and FEMA stepped in after flood disasters to help people rebuild. This unspoken remain-in-place policy reflected the desires of the people who lived in these communities, most of whom had no desire to leave their homes, but it allowed residents of flood-prone areas to avoid bearing the burden of the risk they were incurring. If riverfront towns had had to finance their own levees and self-fund their own disaster recovery, their leaders might have thought twice about building next to the water, but that had never been the case. For as long as FEMA and the Army Corps underwrote the cost of living in risky places, whether through flood protection or post-disaster aid, vulnerable communities had both the incentive and the ability to stay put.

But as time had gone on, maintaining the status quo had become very expensive for the federal government. Not only were flood events becoming more frequent and more costly, but home developers were also pushing farther into the flood zones. This created a vicious cycle of construction, destruction, reconstruction, and re-destruction: the levees and artificial beaches built by the Corps often survived for a mere ten or twenty years before they needed to be repaired or replaced, and disaster-relief officials often found themselves doling out millions of dollars to rebuild communities that were destroyed again a few years later; in the worst cases, FEMA paid to rebuild the same homes as many as four or five times, subsidizing stubborn homeowners to stay where they were. Over time, the mounting toll of flood disasters led the government to shift away from a unilateral commitment to rebuilding in place. The alternative was "managed retreat," or a government-sponsored withdrawal from the most vulnerable places.

This shift began in the 1960s with the creation of the National Flood Insurance Program, a government backstop that offered subsidized flood coverage, filling a market that private companies had long since abandoned. The idea behind the program was to create a stable risk pool by having the whole country pay into a single pot, so that premiums from Texas would pay out to claimants in Iowa one year and vice versa the next. To ensure the risk pool was large enough to break even, Congress made insurance mandatory for most homes built in flood zones. The point was not only to protect homeowners from financial ruin by ensuring they had insurance but also to discourage them from living in flood zones at all: if you didn't want to pay for insurance, you wouldn't move into a flood zone. Or so the theory went.

The program achieved its first goal, at least for a while, but it failed at the second. Because Congress mandated that insurance be offered at "reasonable" rates, most households paid far less than they should have paid for insurance given how risky their homes were. Furthermore, program officials had little power to enforce the floodplain construction requirement: if administrators discovered that a town was building in dangerous areas, all they could do was issue a warning. These shortcomings soon led to a rise in the number of "repetitive loss properties" that filed multiple insurance claims over a given period. Officials also struggled to encourage participation among low-income households like the ones in Lincoln City, many of whom could not afford to buy insurance even at subsidized prices.

In 1993, after a series of monster storm seasons in the Midwest, Congress waved the white flag. Instead of requiring FEMA to help communities rebuild where they were, lawmakers changed the rules to allow the agency to spend a small portion of its budget on mitigating future flood damage. Unbelievable as it may seem, this was the first time the agency was allowed to spend money on preventing future disaster losses. This new funding soon trickled down to officials like Dail, allowing local governments to elevate houses and pursue flood protection projects as they saw fit. The money could also be used to purchase properties that were at perennial risk and knock them down, ensuring they would never flood again. Buyouts offered many towns and cities a permanent solution to what had become a permanent problem.

By the time Fran struck in 1996, Lincoln City had gone decades without a major flood, but it seemed clear to Dail nonetheless that the neighborhood was beyond saving. The homes below Lincoln Street were several feet below flood level, and neither the city nor the county could afford to build a wall around them; elevating the homes, meanwhile, would cost more than the homes themselves were worth. Furthermore, the Corps had abandoned its plans for a second Neuse dam near Kinston. The neighborhood was destined to flood again, and the only way to stop it from flooding was to raze it. Buyouts were the only option.

After Dail and his colleagues at the city finished the mountain of FEMA paperwork required for the buyout application, the county scheduled a community meeting at a local recreation center. Residents from Lincoln City, many of whom had been displaced for weeks or months after the storm, were invited to learn about how the government would pay them to move to newer, safer houses on higher ground. Almost as soon as the meeting began, though, the plan hit a roadblock. The FEMA grant program mandated that all buyouts had to be voluntary, but the residents did not want to leave. When Dail finished his presentation, locals stood up one after the other to attack the idea.

"This is my home," Dail recalled one resident saying. "My great-grandmother lived in this home, it's been in my family for generations." The woman said she wasn't going to leave her neighborhood just because the government had decided to offer her a check. Even those whose houses had suffered extensive damage said they wanted to come back and fix their properties as soon as they could, and asked what the city and county were doing to speed the recovery. The residents also argued that the FEMA money was not enough to entice them to move. If a homeowner still owed a large portion of their mortgage, they said, the buyout would not suffice to pay off the mortgage and make a down payment on another home, especially because the new home would almost certainly be more expensive than the one they were leaving behind.

A few dozen residents did sign up, but most of them lost interest over the following year as the bureaucratic process dragged on. Neither the city nor the county had ever undertaken a grant application as large as this one, and officials soon found themselves lost in a labyrinth of bureaucracy. It took the county almost sixteen months to complete the first handful of buyouts in Lincoln City, and by that time many of those who had considered leaving at first had changed their minds and decided to stay put. Implicit in their decision to stay was the assumption that Hurricane Fran was a freak event, the kind of storm that came once or twice in a lifetime. It may seem counterintuitive, but disasters tend to make people more attached to their homes, not less: rather than taking Fran as a sign that they needed to leave, the residents of Lincoln City determined to hold on to their homes for as long as they could.

As it turned out, that was not very long. A mere three years later, after weeks of heavy rain, meteorologists started tracking a monster storm called Floyd as it made its way up the Atlantic coast. "People had better worry about this one," Dail told the *Kinston Free Press* on September 15 as the storm approached. "The ground won't take much more"—the soil in North Carolina was already saturated with water from another hurricane earlier that month, and any additional rain would have nowhere to go. Dail appeared on the front page of the *Free Press* almost every day that week, with photos showing him posed next to the Neuse or on the phone in his office. The entire population of Lincoln City evacuated the neighborhood as the storm approached, leaving the streets empty to avoid the water. "It's going to be Fran again," Dail told the paper.

It was far worse than Fran.

The next day, the eye of the storm passed twenty miles east of Lincoln City, flooding roads all over the area—"Your paper isn't long enough to list all the roads that are blocked," Dail told a local reporter—and soaking the ground with another fifteen inches of rain. The reservoirs that fed into the Neuse River were already close to capacity, and soon the dam managers had no choice but to open the floodgates or else risk the reservoir overtopping. The next day, as the Army Corps of Engineers released millions of gallons from its Falls Lake Dam near Raleigh, the already swollen rivers in the eastern part of the state climbed several feet higher, rising at a rate of a few inches an hour. Within a few days the Neuse had overtaken three of Lenoir County's seven bridges. The Peachtree Wastewater Treatment Plant, located half a mile from Lincoln City, went underwater and started leaking fecal matter into the river. The deadly cocktail of waste and poison sat stagnant over the eastern flank of the state for ten full days, soaking thousands of houses and seeping into drainage pipes and sewage lines.

If Hurricane Irma brought the full force of nature against the man-made society of the Florida Keys, then Floyd turned man-made society against itself. As the Neuse crested its banks, swollen with rain, it dredged up waste and detritus from hog farms, soybean fields, industrial waste sites, and municipal sewage systems. The water that surged over the eastern part of the state was thick with every poison that had sustained North Carolina's modern economy—gasoline, fertilizer, insecticides, human feces, and runoff from factories. Bloated corpses of hogs, dogs, turkeys, and chickens tumbled along with the current. Later, when rescue workers waded through the ebbing water in search of survivors, they had to smear Vicks VapoRub under their noses to avoid vomiting.

Even before the water had drained out of Lincoln City, the end of the neighborhood had begun to take shape. A week after the storm, Dail was in his office trying to coordinate with other emergency officials. There was a knock on the door, and an elderly couple walked in. They owned a home by the river in Lincoln City and wanted to know where they could sign up for a buyout. As Dail asked them about how much water had gotten into their home (about five feet) and about whether they had flood insurance (they didn't), the wife broke down and started to cry.

"We've got to start over, don't we?" she said, burying her face in her hands.

Over the following weeks, Dail fielded dozens more calls from other families who had changed their minds about leaving Lincoln City and decided that the risks of staying outweighed the pain of giving up their homes. He knew that if the residents were offered another round of buyouts, many of them would say yes. The neighborhood had been destroyed twice in the space of three years. How many people would want to stick around for the third time?

"With Fran, [we] had to go out and walk the streets and knock on doors and send out letters and that sort of thing in order to get people to sign up, because there was not any threat of flooding at that time," he reflected later. "With Floyd, we really didn't have to do that much."

At the time, the new FEMA program was still in its infancy. Officials had only deployed the funds to raze individual blocks or small residential pockets, not entire communities. But Dail and his counterparts in the city planning department were not interested in a piecemeal retreat. Fran and Floyd had shown that the neighborhood was too dangerous to leave alone, and the homes in the neighborhood were worth so little that there was no way to justify a large flood protection project. The whole place had to be abandoned, left for the river to reclaim.

The city of Kinston and Lenoir County hosted yet another meeting at the recreation center, pitching residents again on the idea of a buyout. As the residents stood up to ask questions about the process, Dail noticed that most of them seemed more resigned than they had after Fran. Their anger about the prospect of a buyout in Lincoln City had vanished and had been replaced by what seemed to Dail like a preemptive sense of mourning, a kind of anticipatory nostalgia for the homes they had not yet left. Even though the buyout process would take years to execute, the residents knew what the ultimate result would be. Their neighborhood was not being bailed out. It was being phased out.

III.

The Bargain

M ere days after the river crested, with bands of rain still drizzling over North Carolina, some residents of Lincoln City waded through foul water to see if they could rescue any of their belongings. Those who succeeded in entering their waterlogged homes found themselves wishing they hadn't bothered: the few objects that remained intact in their kitchens and bedrooms were contaminated with all manner of spores and toxins. The flood victims scattered around the county just as they had after Hurricane Fran, some staying with relatives and some taking shelter in FEMA-provided trailers that were clustered in the parking lots of nearby towns. The Democratic primary for mayor took place a few weeks later, having been delayed twice already by the storm, but turnout was still an abysmal 19 percent: no one was around to vote.

As the refugees from Lincoln City waited to return, the buyout program began to take shape. First the city council barred construction activity in flooded areas, and then FEMA handed the state a large chunk of money to buy out Floyd-flooded homes; the state legislature, meanwhile, provided extra funding to help buyout participants relocate to new houses in the same area where they had once lived. Both the city of Kinston and Lenoir County grabbed some money from those grants, and by the middle of December, city and county officials were already talking about the buyout in Lincoln City as a done deal. The unspoken assumption shared by Dail and his fellow government employees was that, even if the residents of the neighborhood didn't want to leave, they certainly didn't want to *stay*.

Among former residents, though, memories of the buyout process are more nuanced. Some recall that the officials from the city gave them a clear choice about whether to take the buyout or not; some say the sheriff came to their doors and told them to get out. Others remember city lawyers cajoling them, strong-arming them, doing whatever they could to clear as many people as possible out of the neighborhood. Some residents say the county gave them a few months to fix up their homes before they asked them to decide; others say the county pressed them for an answer within weeks of the storm.

As always, the truth is somewhere in between all these accounts. The city and county had a statutory responsibility to give residents a choice about whether to take the buyouts, but many residents' homes were damaged beyond repair, and the city wasn't offering them money to rebuild. The residents were traumatized, dislocated, and mistrustful of the local government, and history had given them every reason to be. In the two decades since Floyd, many of these original residents have died, and even more of their descendants have moved across the country or fallen off the map, vanishing in a haze of unknown addresses. Like so much else in Lincoln City's history, the full story of the buyout may be lost for good.

As time went on, however, a kind of snowball effect developed among the displaced households. For the residents closest to the water, whose homes had suffered catastrophic damage twice in three years, taking the buyout was a foregone conclusion. Once those residents decided to leave, their less-damaged neighbors followed close behind. Even for those whose homes had been spared the worst of the flooding, living in a half-empty neighborhood was far from appealing. After all, Lincoln City was made up of people, not buildings.

There was another reality hanging over the homeowners of Lincoln City, one that made leaving the neighborhood seem at once necessary and unbearable: most people in the neighborhood were old. In the years before the flood, many young people had moved away to bigger cities. The neighborhood was still vibrant, but the population was far more skewed toward retirees, people who had raised children in the neighborhood and stayed there after their kids left. Many of these elderly people had been the first in the history of their family to own property, and almost nothing could have made them part from their homes. In the aftermath of the storm, though, they soon found that there was no way they could rebuild. They had paid off their mortgages, but none of them had flood insurance, and because they were retired, they had no income that could help them pay for repairs.

For some elderly people, then, the offer from FEMA seemed like salvation. One of those people was William Lawson, who was seventy-five years old when Floyd struck. Lawson had seen the neighborhood flood every few years since he was a boy, but he knew there was something different about Floyd. The water had reached ten feet on University Street, damaging all three of the Lawson family houses, and the city soon sent them letters condemning the homes because of mold. When they heard about the buyout, they saw an opportunity to accept. He and his son both took the offer, letting go of the little block that had held their whole family history.

When I reached Lawson over the phone, twenty years after the buyout, I asked him how he felt when he looked back on the move with the benefit of two decades' hindsight.

"I'm grateful," he told me. "It was better that we went, it let us move on to something better. It opened things up for us." Lawson had always known the area was vulnerable, and the new house he found on the north side of Kinston was sturdier than the clapboard one his father had built. The buyout meant letting go of the old neighborhood, but it also made possible a safer and more prosperous life, one that he could never have attained were it not for the FEMA money. Many of his neighbors from Lincoln City had ended up moving to the north side of Kinston, too, so it didn't feel like he had lost everything.

Lawson and his family still found ways to keep the past close. His old house was right across the street from a prominent church, so the city never barricaded his block, and city workers still mowed the lawn on the lot where his house used to be. The family's old pecan tree still towered over the space where the house had been. Lawson told me he still went down with his son every summer to gather the fallen pecans.

"We'll still go down there and hang out," he said, "or I'll go down and sit in my car and look around. I'm still close by, so I can feel connected to it. It's better up here, it's safer."

Lawson got a clean ending. For others, the buyout was far more painful.

One of those people was Elwanda Ingram, who was fifty-two years old when Floyd struck in 1999. Although Elwanda had grown up in Lincoln City as a child, she did not live there at the time of the hurricane; a few years after graduating college she had moved to Winston-Salem to teach English at the local university, where she remained for the rest of her career. When she finally hung it up, the campus newspaper ran a photo of her in a shirt that read THE LEGEND HAS RETIRED.

At the time of the storm, Elwanda's parents occupied a small house on University Street, just down the road from Lawson's property. Her father, a stonemason, had built the brick house himself when he first moved to the neighborhood after World War II. It had three bedrooms and a large living room with a fireplace, which had been more than enough room for Elwanda and her sister when they were growing up; in the back of the house there was a neighborhood-standard garden with beans, corn, and collards, all tended by their mother.

Like everyone else on University Street, the Ingrams had always known the neighborhood flooded. Elwanda recalled that once after a storm the streets were overrun with thousands of frogs, so that you couldn't even walk down the street without stepping on one. But neither Elwanda nor her parents had ever seen the water rise high enough to flood the house itself, so the family had never thought to buy flood insurance. After evacuating the neighborhood during the storm, Elwanda's parents returned to Lincoln City to find their home rendered uninhabitable by the flood—the floor had been chewed through by mold, the kitchen had gone black beneath standing water, and countless old family pictures had been blotted out beyond recognition.

Elwanda left Winston-Salem and spent most of the next year living near what had been Lincoln City, helping her parents through the recovery process. After the storm they moved next door to a house owned by relatives, but were only able to stay there for a few weeks before the city shut the power off, forcing them out. This move coincided with the deterioration of Elwanda's mother: she had been experiencing mental health issues before the hurricane, but the separation from the family house and then from Lincoln City proved too much to bear. At the suggestion of her doctors, she checked herself into a mental institution for a few months.

With her mother hospitalized and her father mourning his house, it fell to Elwanda and her sister, Geraldine, to handle the buyout decision. One of Dail's colleagues informed them that the flood damage had rendered the home uninhabitable, which meant that the family could not return there unless they tore it down and rebuilt. The family's best option, the official said, was to take the FEMA money and move somewhere else. The paperwork the official gave them said the buyout was voluntary, but the county presented it as a matter of necessity. The neighborhood was destined to flood again, all their neighbors were leaving, and they didn't have the insurance they needed to rebuild. What choice did they have?

"I don't know if [my parents] wanted to move, but I don't think they had a choice," Elwanda said. "[The county] came in and said, 'You've got to go.'"

Like many former residents of Lincoln City, Elwanda views the flood and the buyout that followed as the result of human decisions, not natural forces. She places much of the blame for the destruction of the neighborhood with the Army Corps of Engineers, which had rushed to build a dam in front of Raleigh but neglected to build one farther downstream. There also seemed to be a kind of racist logic to the buyout itself, one that determined where and when the government saw fit to rebuild. In cities and towns across the state, post-Floyd buyouts tended to take place in the lowest-income communities, places where property values were low and where more residents tended to be Black. Meanwhile, on the ritzy shoreline of the Outer Banks, the all-white beach towns received ample money to build seawalls and restore their eroded beaches.

Though it had an inequitable outcome, this dramatic disparity was caused by simple math. When a federal government agency like the Corps undertakes a flood protection project, it must first justify the project by arguing that the economic benefits outweigh the costs. A hypothetical levee might cost $10 million to build, but if it prevents $50 million in future property damages, the expense is worth it. The denser and more valuable a community's housing stock is, then, the easier it is for the government to justify protecting that community; by contrast, major capital projects in poor and rural communities almost never make financial sense. In Lincoln City, where most homes were worth under $100,000, the cost of building a riverside levee or elevating homes would have exceeded the total cost of all the homes that were at risk. The government needed to make sacrifices to reduce the cost of flooding, and the residents of Lincoln City were the first in line to be sacrificed.

Just before Elwanda's mother checked out of the mental institution, she and Elwanda's father found a new home on the north side of the city, in what was once an all-white neighborhood. The couple had owned their home outright, so they took the entire buyout check and used it as a down payment on their new house. This house was a little newer and a little better appointed than the old house in Lincoln City, which made it more convenient for two people of retirement age, but Elwanda knew it was hard for her parents to let go of their old home. "It has three bedrooms, and it has central air, but otherwise, I will *never* say that it was better than the house my father built," she told me.

Even harder than giving up on the old house was watching over the years as the old neighborhood began to deteriorate: houses fell apart, weeds

took over clean-cut lawns, and people started dum,
borhood rather than taking it to the dump. When far.
took the buyout, they assumed they were just moving out
hood, not leaving it to die. "I don't think people realized tha
going to let it become a jungle," she said.

One day, a few years after the storm, Elwanda and Geraldine went down
to the old neighborhood. As they drove south past Lincoln Street, into a
streetscape that had already become overgrown, they started to smell smoke.

"What's that?" Elwanda asked her sister.

"Smells like they're burning up a house," Geraldine said.

The two sisters followed the smell to the corner of University and Tri-
anon Streets, where their house had been. Their childhood home was up in
flames, surrounded by firefighters from the local fire department. This controlled fire was Dail's idea: once the county took possession of the homes, it
could do whatever it wanted with them. He invited emergency departments
from around the country to come down to North Carolina and practice
their firefighting skills on the real thing.

Elwanda and Geraldine rushed to the new house and picked up their
father, and the three of them drove back down together to watch the blaze.
They sat on the stoop of a neighbor's house, which hadn't been torn down
yet, and watched as the fire destroyed Elwanda's bedroom, then her sister's
bedroom, then the living room. The firefighters had set up a camera to tape
the blaze, and when it was done Elwanda asked them to send her the video.

Elwanda's father was silent as he watched the house burn down. He
had laid the bricks himself and built the walls from plaster of paris, sparing
no expense to ensure that the home would be safe from the waters of the
Neuse. As the firefighters circled the property, they ribbed each other about
Elwanda's father's handiwork—"This is a tough home to burn down," one
of them said. The old brick house, like the neighborhood itself, refused
to die.

IV.

We Lived Here

From the very beginning, Kinston's city planners knew that the buyout would succeed or fail based on what happened to the families who left Lincoln City. The storm had displaced almost one-tenth of Kinston's population, and if even a portion of those people left for good, it would mean a catastrophic drop in tax revenues.

The need to retain the residents of Lincoln City was made even more urgent by the fact that Kinston, like the rest of eastern North Carolina, had fallen on tough economic times. The tobacco factories that once propped up the local economy had closed over the previous decade, and a large T-shirt factory was on its way out, leaving chain stores and healthcare facilities as the dominant local employers. Then, a few months after the storm, DuPont announced that it would lay off more than 1,200 employees from the polyester plant outside the city, giving many residents less reason than ever to stay in the area.

To keep displaced residents from leaving town, the buyout managers offered each of them a $10,000 stipend if they moved into a new home inside the city limits. Fortunately, the area's economic downturn had also created a loose local housing market, and there were plenty of homes for sale in nicer parts of town that once would have been impossible for someone from Lincoln City to afford. Over the five years after the storm, the population of the old neighborhood spread out across Kinston in a kind of miniature diaspora. Some residents filtered into the middle-class Black neighborhoods of the city's east side, while others moved into brick homes on tree-lined streets that had once been open only to whites.

When the city planners had filed away the last of the paperwork, they declared the buyout a success, and from some angles it certainly looked like one. More than 97 percent of floodplain households had accepted the buy-out offer, and around 90 percent of those households had relocated within Kinston. The city and the county had prevented millions of dollars of future flood damage, and because residents' new homes were more valuable than their old ones, the relocation led to a net increase in property tax revenue for the city. In a retrospective on the buyout, the current Kinston city planner wrote that the city's response "serves as an inspiration for other communities facing similar situations."

Yet, when I spoke with Danny de Vries, a geographer who wrote a dissertation about Lincoln City, he took a stern tone.

"I'm curious what your goal is here," he said. "If you're trying to say this was a success story, and this is how things can be done to relocate an entire community, then I'm not so sure. From a governmental perspective of trying to get people out, it was a success; but from a human and cultural standpoint, it was a tragedy."

In the years that followed Floyd, de Vries conducted dozens of interviews with city leaders and Lincoln City residents, many of whom have since passed away. De Vries found that while the government framed the buyout as a measure that enjoyed unanimous support—"[Residents] looked at each other like, this is too good to be true," he quoted one official saying—the reality was far more complicated.

De Vries conducted a phone survey of eighty-six former Lincoln City residents, polling them on their thoughts about the buyout. More than a third of those residents said they would have stayed and rebuilt if given the chance, and the same proportion felt that the buyout had not been voluntary. Only 20 percent, meanwhile, said they remembered being provided with choices other than a buyout, and those "choices" included "get nothing" and "sue the city." More than 40 percent of residents told de Vries they had some amount of opposition to the program; when de Vries visited one participant in her new home, she told him that there was "a lot of bitterness."

Even those residents who had been enthusiastic about the buyout found that they were not as secure as they had believed they would be. Many of the families who took the buyout had already paid off their homes in Lincoln City, but the FEMA money wasn't sufficient to purchase a new home outright, so they had to take out new mortgages. For some families, these new mortgage payments kicked in just as they lost their jobs at DuPont or entered retirement. Not only did they have to pay off the difference in value between their old home and their new one, they also had to keep up with higher property taxes and steep electric bills. In years to come, as the region's economy tanked even further during the Great Recession, the planners who had supervised the buyout noticed a disturbing trend: many of the households that had taken a buyout were now entering foreclosure, indicating that in the long run the buyout money had not been sufficient to integrate them into the higher-income neighborhoods they now occupied. Lincoln City had been the place where many residents had first achieved financial stability, and now that the neighborhood was gone, they found they didn't have the ability to make ends meet in their new homes.

The purpose of the FEMA program was to remove residential properties from floodplains, not to relocate vulnerable populations, so neither the city nor the county ever kept track of what happened to the households that entered foreclosure. However, several former residents of Lincoln City told me that most of them either rented wherever they could find an affordable apartment or bought a trailer home and parked it on cheaper land outside of town. When that wasn't possible, they stayed with family members, even if that meant packing up and moving across the country.

The buyout had begun as an accounting transaction—spend money now, save money later—but the simple act of paying people to leave their homes had unintended consequences. Not only did the buyout destroy one of North Carolina's oldest and most historic neighborhoods, it also destabilized hundreds of lives. By separating the families of Lincoln City from the neighborhood that had held them together, Dail and his colleagues abandoned them to the whims of a market that had no notion of heritage or belonging. As the years went on, that market churned many of them outward, away from Lincoln City. Some were pushed out by financial pressures, others pulled to another state by a spouse or a job or a death in the family, but either way the departures felt inevitable. Once the roots were gone, it was only a matter of time.

—

For Eartha Mumford, though, the destruction of Lincoln City brought her back to Kinston, and closer than ever to the community that had existed there.

Eartha grew up on University Street in a small shingled home by the segregated city cemetery. When the water of the Neuse rose too high, her grandmother would come in from her farm out in the country and drive the family out of town until the flood had subsided. Like many of her high school classmates, she left the neighborhood after she graduated high school, moving to Boston to take a job in the Secret Service. It was in Boston that she met her husband and gave birth to her daughter, but she never felt like she had left Lincoln City behind.

Eartha's husband passed away in early 1999, and Hurricane Floyd struck just a few months later, drawing Eartha's attention back to Lincoln City. When she flew down to see what had become of the old neighborhood, she felt a voice telling her that she needed to come back for good, needed to be closer to her family. She applied for a Secret Service transfer to Virginia, stayed there for about a year, and then moved with her young daughter down to La Grange, just a few miles west of Kinston. She watched over Lincoln City as the neighborhood deteriorated, driving in past the barriers every few months to check on the abandoned streets.

In 2006, Eartha's sister heard about a huge high school reunion that had taken place a few counties over. She told Eartha about it, and Eartha wondered whether the residents of Lincoln City could do something similar for their old neighborhood. The flood had scattered her old neighbors across the country, but many of them kept in touch with one another, and most people hadn't moved very far away. Eartha applied for an event permit from the city and let news of the reunion travel through the grapevine.

Eartha is a loud, imposing woman, one who gets what she wants most, if not all, of the time; she dyes her hair a different shade of pink or purple every month and wears extravagant floral dresses around the house even on Sundays. Even though Eartha had spent most of her life somewhere other than Lincoln City, nobody questioned her when she came forward with the idea for the reunion, nor when she positioned herself as someone who could bring the fractured community back together. Everyone went along with it, either because they respected her tenacity or because they couldn't pass up a chance to return home.

The following Memorial Day, almost a thousand people gathered on the grounds of the Georgia K. Battle Center, a defunct building on Lincoln Street that had once hosted the neighborhood bridge club and other community groups. People came from New York, Arizona, and Alaska. It wasn't just the old generation, either, but also their children, their grandchildren, their in-laws, their friends, and their new neighbors. Each family set up their own tent, played their own music, cooked their own food, so that the grounds became a mishmash revival of the old Lincoln City streetscape, everyone down the street from everyone else.

The reunion went on for three days. When the families took down their tents and packed up to drive home, they donned the T-shirts that Eartha had ordered for the occasion: LINCOLN CITY REUNION . . . WHAT TORE US APART BROUGHT US TOGETHER. The following year, Eartha moved the reunion to Holloway Park, a larger venue that had always been a gathering place for neighborhood kids; when she sent out invitations to the original neighborhood families, she included tickets that bore each family's old address. Soon annual attendance swelled to three thousand, then five thousand, then seven thousand, with people showing up who had heard about

Lincoln City only through stories and rumors. The displaced families, meanwhile, brought old photographs and documents along to the reunion, allowing Eartha to construct history boards that illuminated the story of the neighborhood's founding. To her surprise, Eartha discovered that many of her former neighbors had not known about Lincoln Barnett and the other founders of the neighborhood.

As the reunion blossomed into a major event, it began to provide the city of Kinston with a much-needed boost to its lagging tourism industry, eventually eclipsing the city's renowned barbecue festival. As out-of-towners flocked to the reunion from across the state and beyond, staying for a few days to eat their fill and catch up with old friends, they filled up long-vacant hotel rooms and spent money in downtown restaurants. It was in the city's best interests to keep the reunion going, and Eartha believed she should be able to ask for something in return.

Eartha's request was simple: she wanted to erect a monument to the old neighborhood. It didn't much matter to her what shape the monument took—she thought it could be a pavilion, or a sculpture, or even just a plaque. The idea that got the most traction was for every former resident of Lincoln City to buy a single brick and inscribe their name on it. The bricks would then be joined into a sculpture that would sit at the entrance to the former neighborhood. Not only would the sculpture serve as a testament to the lasting unity of the community, it would also pay homage to the brick homes that the early residents of the neighborhood had built.

The city council told her that such a monument was impossible. One condition of the FEMA buyout program, they said, was that the city could never build anything new on the buyout tract.

"When it came to Lincoln City, they always bumped me," Eartha told me. "When they saw me coming, they [were] like, 'Uh-oh, here she comes,' and they would stop the city hall meetings when I would walk in. They just tried to get rid of me."

The city's reasoning was flawed. FEMA regulations prohibit most new development in buyout areas, but they don't prohibit monuments or other memorials. A few towns and cities that have conducted FEMA buyouts have transformed their buyout sites into parks, greenways, or wildlife refuges, and some have erected small monuments to the communities that once existed there. Still, Eartha needed the city's approval to move forward, and over the years the council made it clear to her that honoring Lincoln City wasn't a priority. She got the message and eventually gave up.

A few years later, in January of 2021, I met Eartha in the parking lot of Holloway Park. The pandemic had forced the community to call off the gathering that year and the year before, but as we talked, Eartha sketched for me what the reunion looked like in normal times. Tents would extend for hundreds of feet, filling the faded basketball courts and the vast expanse of grass beyond that. The younger generations played on the playground or chased each other around the pavilion, while the old folks stood in line for ribs, cornbread, and collard greens. People pitched folding chairs and sat in small groups, catching up on each other's life and reminiscing about a time and a place when life was simple, when they had wanted for less.

I asked her if she worried about the Lincoln City community drifting apart during the pandemic, without a reunion to bring everyone together. Eartha said she didn't believe that would happen, in part because the bonds of the neighborhood were so strong and in part because so much had already been lost. Over the years the reunion had become less a way to keep people together and more a chance for people to mourn the place they once knew.

"Everybody had a love for Lincoln City, and they still have a love for Lincoln City," she told me, "but they know that Lincoln City is no more."

V.

What Could Have Been

L incoln City is haunted by a counterfactual. What would have happened if there had been no buyout, if the neighborhood had been able to rebuild? This is a tormenting question because there's no way to know for sure. We can never be certain how many people would have moved back to Lincoln City if given the chance, nor what the fate of the rebuilt neighborhood would have been.

There is, however, another town nearby that represents a kind of control group for the buyout experiment. As Hurricane Floyd swept through North Carolina and ravaged Lincoln City, it also brought severe flooding to a historic town called Princeville, another struggling but self-sufficient community, about fifty miles away. In Princeville, too, FEMA offered a large-scale buyout, offering to move everyone out and let the town return to the water. In Princeville, though, the residents said no.

Granted, Princeville's situation was somewhat different from that of Lincoln City. Founded in 1885, it is the oldest town in the United States to be chartered by Black people, and as such it has an outsize historical importance for a village of five thousand people. This unique legacy meant the town received a flurry of national attention after Floyd: a veritable troupe of celebrities and political figures including Al Sharpton, Jesse Jackson, Dick Gregory, Tipper Gore, Bill Clinton, and the musician Prince all flew in to offer their support or sent donations from afar. The town was also an autonomous jurisdiction, rather than a neighborhood like Lincoln City, which meant that the town council had the authority to accept or reject FEMA's buyout offer. After weeks of agonizing debate between federal government engineers and advocates like Sharpton and Jackson, the council voted three to two to rebuild.

Like Lincoln City, Princeville had always had a flooding problem. The freedmen who founded the town had done so on land granted to them by white plantation owners on the other side of the sluggish Tar River, land that was so low and prone to water that the planters had no use for it. In the 1960s, after the town saw its tenth flood in a hundred years, the Army Corps of Engineers erected a twenty-foot levee around the town's perimeter. The levee protected the residents from flooding for decades, but Floyd exposed serious problems with its structure, problems that would jeopardize the town for years to come. It was a turn of events that foreshadowed the tragedy of Hurricane Katrina six years later, when the Corps's levee system around New Orleans failed.

The town council of Princeville rejected the buyout, believing that the Corps would repair the flaws in the levee before the next storm. In the aftermath of an apocalyptic flood, with the national spotlight trained on the town and President Clinton promising immediate relief, this belief seemed more than justified. But as the years went on, the levee project stalled out in an endless sequence of meetings and reports. Meanwhile, the residents rebuilt on the only land they had ever known, erecting a new town hall and hundreds of new homes.

In the summer of 2016, the Corps presented the town with a final scheme to fix the levee, but the plan came too late. A few months later, a powerful storm called Matthew barreled up the Atlantic coast and dumped several feet of rain on eastern North Carolina. The Tar River crested its banks and broke through the levee in exactly the ways the Corps had predicted it would. The town filled up with water, the residents evacuated, and the whole process began anew.

FEMA once again offered to buy out the whole town, and the council once again rejected it, but this time Princeville's leaders were only prolonging the town's demise. The few stores and shops that had managed to rebuild after Floyd had no money to do so after Matthew, which left the town with no grocery store, no laundromat, no restaurants. Dozens of people left the town of their own accord, and some individual households finagled FEMA buyouts on their own. The town is still standing today, but it is a shell of its former self. Dozens of homes stand empty, scarred with spots of mold left over from the last flood, and the historic center is a maze of empty lots.

In retrospect, the moment that fixed this ending was not the moment that the town rejected the buyout, but the moment that the white landowners on the other side of the Tar River ceded a flood-prone tract of land to the freedmen who wanted to start their own community. The prime culprit in Princeville's destruction, as in Lincoln City's, is not nature or even climate change, but the many-stranded racism of American society, the silent hierarchy that divides land along lines of race and class.

Lincoln City and Princeville took opposite paths in the aftermath of Hurricane Floyd, but both paths ended with the towns emptied out, their residents scattered, their histories half-erased. From the very beginning, places like these were never seen as worth protecting. They got a broken levee, or no levee at all, and their inhabitants were left to bear the burden of a risk that white people all around them never had to confront. When the status quo of climate change became untenable, it was these places that were sacrificed, these places that were the first victims of the logic of retreat. The rivers that run through the two towns may start far apart from each other, but they empty into the same ocean.

CHAPTER THREE

Burnout

WILDFIRES, INSURANCE,

AND THE HOUSING CRISIS

Santa Rosa, California

The Fire on Tubbs Lane

E ven five years later, the precise facts of the Tubbs Fire incident remain
unknown. Here is what we can say for sure.

Early on the evening of October 8, 2017, the wind started to blow
in Northern California. This wind had begun as a high-pressure system
rotating clockwise over the deserts of Nevada and Utah, but the de-
scending jet stream had given it an extra push, boosting it over the Sierra
Nevada and into California. The desert wind was hot and dry, and once
it crossed the mountains it sought the cool expanse of the Pacific. It
rushed down the mountains and into the rolling hills north of San Francisco, swooping through the Napa and Sonoma Valleys, the most famous
winemaking regions in the United States. The wind reached speeds of
more than eighty miles an hour, equivalent to a Category 1 hurricane. It
roared through the vineyards, shook the trees, kicked up hazy storms of
leaves and grass.

That night was a balmy one in Sonoma County, and the conditions were
ideal for a fire—the humidity was around 10 percent, indicating an extreme
drought, and the moisture content of the forests around the valley was an
astonishing 3 percent. Earlier that day, state officials had issued a "red flag"
warning for potential fires in the area, but the wind speeds coming out of
the desert had exceeded their expectations. With gusts that high, and so

much flammable vegetation around, it was almost inevitable that something would ignite.

Around nine thirty in the evening, an isolated blaze started at a vineyard property on the outskirts of Calistoga, a town at the northern edge of the narrow Sonoma Valley. The property, situated just off a street called Tubbs Lane, sat on more than ten acres and was big enough to have its own private electrical system. It's unclear what ignited the fire, but investigators have speculated that the high winds may have broken a piece of the property's electrical system, knocking over a pole and sending sparks onto the surrounding vegetation.

By 9:45 p.m., multiple people had reported seeing a fire on the property. The next-door neighbor woke up to the sound of his smoke alarm, by which time the fire was already snaking across his lawn; around the same time, a police officer who was in the area to respond to a separate fire said he saw a "large glow" at the house. A few minutes later, another neighbor described a wall of orange flame heading up the ridge behind the house. When a fireman arrived at the scene shortly after, he found the fire moving to the north, consuming a copse of trees; the fire was moving at a "rapid rate of spread," he reported, and he knew he couldn't put it out himself, so he left the scene and called for reinforcements.

Soon after the fireman left, the fire ran into the desert winds coming out of the northeast. It turned on a dime and headed south, and within a few minutes it had breached a steep range of foothills called the Mayacamas, which separate Sonoma Valley from the city of Santa Rosa. It would be difficult to imagine more flammable material than the vegetation that blanketed the Mayacamas that night: a large rainstorm early in summer had released California from a punishing drought, causing grasses and shrubs to proliferate on the mountain slopes, but in the months since, the drought had resumed, drying out the new vegetation into kindling. The blaze now fed on this mother lode of fuel, swelling in size until it was swallowing hundreds of acres a minute. As the fire climbed higher up into the hills, it encountered even stronger winds, which catapulted it forward onto more vegetation and further accelerated its ascent. It had only been two hours since ignition, but already the fire had become too large and too fast for the local or state firefighters to control. It was not a question of if the fire would reach Santa Rosa, but when.

Vicki and Mark Carrino lived less than ten miles away from Tubbs Lane, but as far as they knew, they didn't have any reason to be worried about a fire in Calistoga. It had been decades since a fire had crested the Mayacamas, and even if such an improbable event were to occur, there was still almost no chance that it would reach the couple's house: there were a half-dozen vineyards that separated their street from the forest, and Vicki had always been told that the hardy grape vines would serve as natural firebreaks, hampering the progress of any rogue blaze. Fire was something that happened out in the deep wilderness, not in exurbs like theirs.

The Carrinos had been out late the previous night at a neighbor's birthday party, and they had spent most of Sunday relaxing on their back deck, gazing out at the dozens of pine and fir trees that encircled their property. After a glass of wine each, they went to sleep at a reasonable hour, around ten o'clock, leaving their cell phones in the other room.

Two hours later, at midnight, the landline rang in the kitchen. Vicki answered it and heard her daughter on the phone: she had called their cell phones a dozen times each, she said. There was a fire coming over the hills, moving thirty miles an hour toward their neighborhood. They needed to get out as fast as they could.

Even with the windows closed in their bedroom, Vicki and Mark could smell the smoke. They walked bleary-eyed to the living room and looked out toward the Mayacamas, where they saw a red glow rising over the hills just a few miles away, silhouetting the pine trees in a sinister tableau. The couple raced around the house grabbing whatever they could think to take—a change of clothes, their passports and, mercifully, their insurance papers. By the time they returned to the living room five minutes later, the blaze was halfway along the downslope, rushing for the vineyards. As they stepped outside, a wind howled around them, so fast and hot that it stung their cheeks. They jumped in their truck and gunned it down their driveway, speeding away from the path of the blaze.

The forward edge of the fire reached their street no more than ten minutes after the Carrinos left their house. As the blaze breached the outskirts of Santa Rosa, it attained an almost baffling explosive power. The combination of high winds and ample fuel had allowed the fire to gather intensity even as it whipped south across the Mayacamas. It blasted through natural and unnatural firebreaks alike, scorching the protective vineyards in a matter of seconds and jumping across wide stretches of asphalt that should have slowed it down. The wind on the downslope hurled embers as far as a mile ahead of the blaze, sparking new isolated fires that the larger fire soon absorbed.

Fires need to consume oxygen to continue burning, but by the time the blaze reached the Carrinos' street it had sucked almost all the oxygen out of the air, draining the entire forest of combustible gas. There was still oxygen inside the homes, though, and as the fire ripped down the mountains it sought out this fuel. The flames rushed through the vents and air ducts of the Carrinos' house and detonated the flammable gas inside, burning the home from the inside out rather than vice versa. The entire structure turned to ash in a matter of seconds—leather, plastic, paper, wood, plaster, and concrete alike vanished into the same vapor, leaving behind only a brick fireplace and the charred outline of the home's foundation. When the Carrinos returned a few days later, they discovered that a bauble of purple amethyst hanging in their doorway had transformed into dark orange citrine, indicating a chemical reaction that can only occur at sustained temperatures of more than a thousand degrees Fahrenheit.

The fire kept going. Sustained by the oxygen bombs in homes like the Carrinos', it fanned out into the nearby hills and entered a wealthy section of the city known as Fountaingrove, stampeding through subdivision after subdivision. The homes it burned were large four-thousand-square-foot mansions, many of them with pools or panoramic back decks. They belonged to the upper tier of Santa Rosa's wealthy—surgeons who worked at the massive hospital down the hill, managers and owners of high-dollar wineries, senior employees at Bay Area tech companies. These families had come to Fountaingrove to find an aerie, an escape from the density of the city below, but now they were paying for their decision to encroach on nature. The fire incinerated millions of dollars every minute. It leapt from cul-de-sac to cul-de-sac, torching the vineyards and stables that lay between them, scattering livestock down the hills. Within hours, the fire had burned more than ten thousand acres and destroyed more than a thousand structures.

Having crossed the mountains, the fire descended the hills and pushed southwest toward central Santa Rosa. If a wildfire in the hills around the city was rare, a wildfire in the city itself was almost inconceivable—Santa Rosa sat outside the high-risk territory known as the wildland-urban interface, where human communities overlap with flammable forest environments. The city's close-cropped lawns and wide asphalt roads were inhospitable to wildfires, and the city was separated from the hills by the long curve of Highway 101, a massive six-lane freeway with an additional two lanes of frontage road on each side. The freeway was the ultimate firebreak, a stretch of impervious material so wide as to be impregnable: a strong blaze might charge right up to the highway, but it would never cross over to the city itself.

That was how the theory went, at least, but all theories are temporary, and in a changing climate their life spans seem to be shorter than ever. No one in Santa Rosa, or in California for that matter, had ever witnessed a fire like Tubbs. As the blaze reached the bottom of the hill, the north winds whipped its embers into long eddies, catapulting them off the hills and onto the asphalt freeway. The embers landed in ashen bursts along the highway, pelting a stream of escaping cars. Like falling arrows in a siege, the embers landed one after the next, igniting lawns and trees and telephone poles, bursting electrical transformers and turning mailboxes into pillars of flame. The fire reconstituted itself on the other side and blazed onward as though it had never encountered the highway at all, incinerating a Taco Bell, a Kmart, and a complex of courtyard apartments along the way.

Kevin Tran had been monitoring the fire on Facebook all night, but even as it reached the bottom of the hills, he and his parents still weren't sure whether they should leave. The Trans lived in Coffey Park, the closest neighborhood to the freeway and one of only a few remaining middle-class areas left in Santa Rosa. No one in the neighborhood had ever heard of a fire reaching the area, so most people weren't even tracking the blaze.

Around one o'clock in the morning, Kevin started to hear a howling wind; he could see a faint orange glow coming from Fountaingrove, but there was no smoke in the air above his house, and the street was empty except for bleary-eyed locals stepping outside to walk their dogs. He and his parents sat in the kitchen and waited until Kevin came across a notice on Google around two in the morning: the city was evacuating Coffey Park and most of the neighborhoods around it. By the time he rushed his parents outside, half the neighborhood had already started their cars and left. Most people still didn't think the fire would reach the neighborhood, but it was better to be safe than sorry.

Kevin's parents had come to the United States from Vietnam before Kevin was born, and to them Coffey Park was the very image of the American dream: the neighborhood was middle-class and diverse, with an active community of walkers in the central neighborhood park and an annual Fourth of July fireworks show. They appreciated the cute street names—Mocha, Sumatra, Cashew—and over the previous summer Kevin had struck up a friendship with a group of other twentysomething guys in the neighborhood who all played the popular augmented-reality game *Pokémon Go*. The home in Coffey Park was the first that Kevin's parents had ever owned—they had previously lived on Section 8 vouchers—and Kevin's mother was hesitant to leave it behind. She told Kevin she wanted to drive out of her neighborhood by herself, following him and his father in a separate car, but Kevin worried she would turn back and retrieve their photo albums from the house, so he made her drive with him. His father followed behind them with the family dog.

The scene as the Trans evacuated was silent, eerie, so strange as to seem almost dreamlike. Even though the entire neighborhood was scrambling to leave at once, the drivers at the front of the line were still obeying stop signs and red lights as they turned out of their cul-de-sacs and onto the main road; the drivers at the back of the line, meanwhile, were honking their horns in fear, since they could see the red glow raging in the hills just a mile away. As he drove out of the neighborhood, Kevin watched the wind whip up a small brush fire into what looked like a flaming tornado.

Just like the Carrinos, the Trans made it out in the nick of time. Just thirty minutes after they left Coffey Park, the first rogue embers started to land on the neighborhood. Trees burned down to stubs, cars crumpled like balled-up sheets of paper, and transformers exploded into the sky like the fireworks Kevin had loved to watch. The narrow lots and communal fences that had lent the neighborhood its charming suburban feel now served as perfect conductors for a blaze that might otherwise have fizzled out: the blaze roared down the length of every fence and buffeted its way into homes like Kevin's, melting them like wax.

The Trans stayed the night at the Vietnamese restaurant they owned in nearby Rohnert Park, but they didn't sleep—Kevin stayed on Facebook all night, looking for an update on their neighborhood. The wind from the hills died down around four in the morning, and it wasn't long after that before the fire ceased its march through Santa Rosa—without the gusts to push them onward, the flames couldn't cross the railroad tracks west of Coffey Park.

Kevin left the restaurant at the break of dawn with his father and sister, who also lived in Coffey Park, hoping they could reach the neighborhood before the National Guard erected barricades and blocked residents from returning. They got off the freeway a few exits before Coffey Park and drove through the silent residential streets to the south of the neighborhood. The tree cover over their heads vanished all at once, followed by the grass on the lawns and the sidewalks around them. They emerged into what seemed like a desert, a vast off-white expanse smeared with the charred remnants of houses and cars. They could see for miles across the neighborhood, but the fire had destroyed all the street signs, so Kevin had to rely on Google Maps to find his way to the family's house on Hopper Avenue. Even after he reached his block, he couldn't figure out which of the burned-out lots was his family's, until he saw the car he used for his job at a Verizon store. The car was the only intact vehicle in the entire neighborhood, and it had only survived thanks to a mistake Kevin's father had made when installing the lawn sprinkler system—the sprinklers sprayed out onto the street instead of onto the grass, and they had kept the car from burning when they turned on overnight.

Kevin and his father got out of the car and stepped into the pile of ash where their house had been. He felt almost numb as he walked through the wreckage, maybe because there *was* no wreckage—nothing had survived except a few small objects blackened beyond recognition.

Then Kevin heard someone calling out to him. He looked up and saw a neighbor walking over to him, through the space where his backyard had been: their houses had abutted each other before the fire. Kevin looked around him. More people were filtering back in, hundreds of them, hoping to get a glimpse of their houses and search for surviving valuables while they still could. These people had worked for years to buy homes in the neighborhood, had stretched their budgets and emptied their savings just to get a foothold there. Kevin's parents were no different. As he walked back and forth around the burned-out yard, his mind kept coming back to the same question that was consuming all the climate refugees around him.

Where were they all going to go?

II.

Less Than Zero

A flood may fill the walls of a home with mold, rupture the floors, and ruin the belongings inside, but when the water recedes it leaves the structure of the home intact. The damage can be repaired. A wildfire shows no such mercy. The Tubbs event erased every trace of Coffey Park, Fountaingrove, and the neighborhoods around them, leaving families like Kevin's with nothing but empty lots. More than five thousand houses disappeared from the city in the span of a few hours, along with the streetscapes and electrical grids that had linked them together. It would be years before new communities could be built to replace the ones that had vanished, and in the meantime the fire victims had nowhere to live.

There are not enough houses in California—this is an incontrovertible fact, and it was true long before the Tubbs Fire ignited in October of 2017. Over the previous fifty years the state's housing shortage had created a complex web of social and economic crises, forcing thousands of people onto the streets even as rapid gentrification tore through cities like San Francisco. There were already millions of people in the state who did not have a home or who were struggling to afford the homes they did have, and the victims of the Tubbs Fire now joined them in a desperate scramble for affordable shelter.

California's population more than doubled between 1960 and 2010, surging from fifteen million to thirty-five million and making the Golden State into the world's fifth-largest economy. The reasons for this meteoric growth were many, but most of them had to do with jobs: the state was home to any number of boom industries, from agriculture to film to tourism, and toward the end of the century the Bay Area gave birth to the tech sector that would come to dominate the national economy. The people who arrived to work in these boom industries needed somewhere to live.

Except the power to build housing lay with the dozens of towns and cities that made up the greater Bay Area, whose city councils and commissions had the authority to approve or deny all new development permits. As Conor Dougherty details in his book *Golden Gates*, these elected officials had every political incentive to deny new housing rather than approve it. The oldest, wealthiest, most engaged constituents all tended to be long-time homeowners rather than newcomers or renters, and these homeowners abhorred new development that might disrupt their idyllic single-family neighborhoods or, even worse, bring low-income people into their midst. "The people who showed up to speak in favor of new projects were the developers who'd proposed them, the trade organizations they paid to shill for them, the unions who worked for them, and the community organizations whose wheels they had successfully greased," Dougherty writes. "The people who showed up to oppose new projects consisted of environmentalists, angry neighbors, and community organizations that felt their wheels hadn't been greased enough." The latter groups tended to vote in local elections, which meant that any city commissioner who wanted to keep her job sided with them.

The result was a classic tragedy of the commons: everyone supported new housing in theory, but no one supported it in practice, and indeed every proposed housing development attracted torrents of hatred and criticism from the homeowners who lived in the area. The locals who opposed new housing became known by the pejorative term NIMBYs, which stands for "not in my backyard"—they didn't care if a developer built an apartment complex in someone else's neighborhood, but they didn't want it built in theirs.

As if this political dynamic were not enough, two idiosyncratic state laws further constrained housing supply. The first law was Proposition 13, a 1978 ballot initiative that froze residential property taxes for the entire state and in turn eliminated incentives to build new housing. Before the law, the owners of a vacant lot would have paid increased property taxes as the value of the lot went up, which might have prompted them to build an apartment complex on the lot and collect rent; with property taxes frozen, though, they had no reason to build. The second law was the California Environmental Quality Act, signed into law in 1970, which was intended to protect citizens from the downstream impacts of development and construction. If a city wanted to build an airstrip or a landfill, for instance, a homeowner could sue to stop the project, forcing the city to conduct a lengthy analysis to ensure that the project wouldn't cause adverse health impacts for the plaintiff and their neighbors. The motivation behind this law was benign, even progressive, but the language of the statute left it open to abuse. The standard for negative environmental impact was low, the review process was time-consuming, and it was far too easy for individuals or neighborhood groups to sue. The result was that thousands of homeowners sued to block affordable housing developments and homeless shelters under the auspices of environmentalism even when their real motivations were less charitable.

Almost every major city in the United States faces a housing shortage of some kind or another, but California's combination of restrictive laws and rapid growth has made homeownership burdensome for middle-class families and all but impossible for low-income households. According to one estimate, the shortage exceeds four million units, which means ten million people need a home but do not have one. Although the housing crisis originated in the meteoric growth of the Bay Area and Los Angeles metroplexes, its tentacles have spread to every corner of the state, raising costs in the agricultural towns of the Central Valley and in remote mountain towns like Greenville.

Santa Rosa is more than an hour away from San Francisco by car, but the region is still well within the reach of the Bay Area's economy, and over the past thirty years it, too, has felt the pain of the region's housing shortage. Napa and Sonoma Valleys had their own growth spurt toward the end of the twentieth century, and just like in San Francisco, the towns and cities in the surrounding area struggled to create adequate housing to keep up with demand.

The primary driver of the region's growth was wine, and in particular wine tourism. In the 1960s, a new generation of winemakers led by Robert Mondavi and Louis Martini descended on California to capitalize on a postwar boom in wine consumption. The climate and soil in the region were ideal for grape cultivation, and this new generation of producers had big plans for the region: they pressed billions of grapes per year, pushed for recognition from prominent juries, and built enormous wineries that attracted thousands of tourists per year for tours and tastings. By the turn of the century the region's economy had become less agricultural than tourist-based, just as dependent on revenue from vacationers as it was on revenue from the sale of actual wine. At the same time, the tech boom minted a new stratum of nouveau riche, many members of which sought second homes and rental properties near famous vineyards. The influx of tourists and vacationers led to an explosion in luxe resorts, boutique hotels, classes for winemaking and wine tasting, and bike tours around the valley. The wineries themselves, meanwhile, required tens of thousands of laborers to work the vineyards each year, and all those laborers needed places to stay as well.

The combined demand for homes in the area far outpaced the total number of homes, creating a bidding war for the region's scant supply of available housing. The wealthiest buyers claimed the largest homes in areas like Fountaingrove, which relegated middle-class families like Kevin Tran's to quasi-suburban neighborhoods like Coffey Park. Affordable housing in Santa Rosa was scarce, which meant that low-income families had almost no chance of getting a mortgage. Having been priced out of the ownership market, these families had no choice but to rent on a permanent basis. Further down the income spectrum, other renters took out leases in courtyard apartment complexes or in small backyard dwellings known as "granny units." The least fortunate had no choice but to sleep in shelters, or in their cars, or in tents, or on the street.

At the time of the Tubbs Fire, the overall housing vacancy rate in Santa Rosa was around 3 percent, one of the lowest rates in the country. Except for a small number of homes that were changing hands or undergoing renovations, there were no unclaimed houses anywhere in the city. The same was true in the surrounding counties. For renters in particular the winemaking area was only becoming less affordable: median wages in Sonoma had declined by 6 percent between 2000 and 2015, but median rents had increased by 15 percent over the same period. In the years before the fire, the number of homeless people in the county reached an all-time high of more than three thousand, filling up the area's meager supply of shelter beds.

The fire made this already dire problem even worse. The blaze destroyed more than five thousand homes in the city of Santa Rosa alone, eliminating a full 5 percent of the city's housing stock. If there had been zero affordable homes in Wine Country before the fire, there were now somehow less than zero. The ten thousand–odd people who had lost their housing in the fire were forced to compete not only against each other but against the rest of the area's unhoused population for access to a minuscule number of available homes. Other fires that ignited on the same night as the Tubbs Fire had destroyed at least two thousand more homes in nearby Napa and Solano Counties, erasing the scant backup housing stock in the areas surrounding the Santa Rosa blast zone. The already frenzied bidding war now collapsed into a state of outright turmoil. Prices jumped 10, 20, 30 percent overnight. The buyers outbid each other by hundreds or thousands of dollars, families drove two hours to view apartments that wouldn't even fit them, displaced homeowners emptied their savings to place deposits on condos they hadn't even seen.

In the thick of this chaos, of course, the wealthiest victims found housing first and fastest. They leveraged disposable income and social connections to grab up vacation homes or empty condominiums before they hit the open market. Many residents of Fountaingrove also owned second homes in the valley that they rented out to vacationers, so those residents moved a few miles down the road while they waited to rebuild. Others leased vacation homes that belonged to their friends and family. Those who couldn't find a home through their personal networks turned instead to a chaotic open market. Doctors, lawyers, and winemakers who lived in Fountaingrove used their ample savings to bludgeon their way into leases that should have been going to middle-class families; the wealthiest families, meanwhile, just bought vacation homes outright with cash, telling themselves they could flip the properties in a few years after the recovery.

Even among higher-income families, though, the scramble for housing was not a fair fight. Many of the wealthiest victims had insurance policies that provided them an almost unlimited tranche of money for "additional living expenses." Such policies required the insurer to pay out rental costs equivalent to the value of the victim's home, which for many people in Fountaingrove meant upward of $10,000 a month. As the bidding war continued, some landlords in Sonoma County began hiking their prices to absurd heights, setting rent at $8,000 a month because they knew that victims out there had insurance policies that would allow them to pay that much or even more.

As the scramble continued over the coming months, the Sonoma housing market tightened even further, so that soon even those whom the fire had spared began to feel the squeeze. Anyone looking for an apartment in Santa Rosa had to compete against thousands of displaced households, many of them armed with an unlimited spout of insurance money. The low-income neighborhoods on the southwest side of the city, which before the fire had offered an affordable palace of refuge to vineyard laborers and their families, now saw a dramatic spike in housing costs. When their leases came up for renewal, many of these families found that their landlords wanted to raise their rent by hundreds or even thousands of dollars. In other cases, the landlords had lost their own homes to the fire and now wanted to live in the second homes they had been leasing out. Anyone who had been homeless or between houses before the fire was now more than likely to find themselves shut out of the market altogether. Shelters and encampments in the area, most of which had been at capacity even before the blaze, now filled up with fire victims who sought temporary beds alongside families who had been unhoused for years. The door had been open just a crack beforehand, but now it had been slammed shut.

There was no room in this frenzy for someone like Henry Arriaga, who had been renting in Coffey Park for five years before the fire. Henry was the project manager for a company that did flooring repairs in residential homes around the area—not glamorous work, but he liked numbers, and he appreciated how the job demanded an eye for detail. It had taken him years to reach the project manager position, and the upgraded salary allowed him to support his wife, Vanessa, and their young daughter, but buying a home was still a pipe dream: even before the fire caused a spike in prices, the bidding wars for homes in Coffey Park were well beyond their means. Their daughter loved the neighborhood, though, so they planned to stay there even if it meant renting for another five years while they saved.

After their rented house burned in the fire, the Arriagas stayed with Vanessa's aunt for the first few weeks, but they soon realized that wouldn't last very long—the house was too cramped with five people living there, and their daughter needed a space of her own in which to recover from the trauma of the evacuation. The couple had paid around $1,800 a month to rent their two-bedroom house in Coffey Park, and it didn't take Vanessa long to realize that there was nothing available in Santa Rosa for even close to that amount. Nevertheless she drove back and forth across the city in search of something comparable to the old house, visiting dozens of houses only to leave empty-handed when the landlord asked her for $2,500 a month. She set her sights lower, focusing on courtyard apartment buildings and backyard dwellings, but there, too, the Arriagas found themselves unable to get a foothold: a landlord might offer a monthly rent that sounded affordable, but he would also ask for two months' rent up front plus a security deposit, which was money the family didn't have. As the holidays approached, Vanessa and Henry started to look outside Santa Rosa, searching farther afield in smaller cities like Petaluma and Rohnert Park.

At last Henry and Vanessa found an apartment they could afford, a triplex unit in the city of Sonoma, about twenty miles away. It was an unfamiliar and cramped environment for their daughter, it more than doubled the length of Henry's commute, and it cost more than $2,000 a month, but at least it was a place to stay. By the time the lease ended a year later, the couple had saved up enough money to rent an apartment back in Santa Rosa, just a mile or so away from Coffey Park, but once *that* lease ended they had no idea what they would do. The couple had taken out renters insurance before the fire, and they had received small assistance payouts for the first eighteen months after the disaster, but now their insurer was cutting them off, saying they had had enough time to find adequate housing. Without the insurance payments they wouldn't be able to afford an apartment in Santa Rosa or even in Sonoma. They started looking at townhomes, mobile homes, modular homes, anything at all if it would fit their budget, but they just couldn't make the math work. The couple had had a good relationship with their pre-fire landlord in Coffey Park, but when the landlord rebuilt the burned property and started renting it out again, he charged far more than Henry and Vanessa could afford. Even two years after the fire, after multiple moves, it felt like they hadn't even begun to work their way back to Coffey Park.

Around the same time, Henry's flooring company started to get calls from fire victims who wanted to rebuild their homes. Once the county cleared the debris and repaved the streets, homeowners who had been lucky enough to receive sufficient payouts from their insurance companies could start designing their replacement houses. Flooring tended to be one of the last things they handled before moving in. As project manager, Henry seldom did site visits, but there were a few occasions when he found himself driving into the Fountaingrove hills to check on the status of a job, or exiting the freeway at Coffey Park to drive through the streets where he had once lived.

It was on one of these site visits that Henry began to realize just how far his family's fate had diverged from the fates of the wealthier victims around him in Santa Rosa. His company was doing a floor job for an enormous house up in the fire-burned hills, in a ritzy neighborhood called Wikiup, and the client happened to be there when Henry stopped by to check on how his contractors were doing. When Henry finished his inspection, he chatted with the client for a few minutes in the driveway.

"You know," said the client, "I'm kind of glad the fire happened." He gestured at his sloping backyard and at the panoramic view of the valley behind it. "I used to have all these trees here," he said, "and they blocked the view. Now they're all gone."

Henry's nails dug into his palms. He felt anger rising inside him, and he knew he would snap if he didn't get away from the client. He muttered something about another job, jumped in his truck, and drove as fast as he could down the hill and back into the valley.

—

For other refugees from Coffey Park, the whirlwind aftermath of the fire pushed them even farther away, out of California altogether.

José Guzman and his family had owned a home in Coffey Park for fifteen years, having come to Sonoma County from the state of Michoacán in central Mexico. He had supported his wife and four children with an assembly line job at the local plastics factory, saving up enough money to take out a mortgage in Coffey Park. The house was small and charming, but nevertheless the mortgage had been difficult for José to maintain, and he had compensated by taking out the narrowest available insurance policy, one that covered just $240,000 in damage. When the home got destroyed in the fire, José found that his claim payout was nowhere close to the value of the property, and furthermore that it didn't include any temporary living expenses—not only could he not afford to rebuild, but he also couldn't afford to rent anywhere else. José had never known that the neighborhood faced such severe fire risk, and now he and his family were paying for it.

In the immediate aftermath of the disaster, José and his family weren't sure if they wanted to stay in Coffey Park or even in Santa Rosa. They liked the area, but José thought of himself as a sensible man, and it seemed foolish to live in a place where you could always see the fire-prone hills. In the end, the firm hand of the housing market made his decision for him. By the time the Guzman family got its bearings, there were no hotels or short-term apartments anywhere in Sonoma County. The landlords José spoke to were asking for five, six, or even ten thousand dollars a month in rent, sums so ludicrous they made him want to slam down the phone. FEMA offered his family a trailer out in Sacramento, but that would have meant a two-hour commute to the plastics factory in each direction. José was willing to do a lot to keep his family afloat, but he wasn't sure he could do that.

The six members of the Guzman family stayed with one of José's stepdaughters in nearby Petaluma for a few months. Just like the Arriagas, they soon found they couldn't live doubled-up in such a small house, but there didn't seem to be any other homes in Sonoma County or anywhere else in California. Their only option was to stay in the closest place where they had family members who would put them up. As it turned out, that place was not close at all—it was Louisville, Kentucky, where José's wife had relatives who offered the family their guest room. José felt absurd telling his children that they had to get in the car and drive them more than two thousand miles across the country just to find a place to stay, but that was the reality. The drive over the Rockies and across the interminable expanse of the Great Plains took almost a week, and the family was quiet most of the way, packed into the car with all their belongings. They didn't know anything about the place they were going, and they had no idea how long it would be before they could make their way back home.

José figured he would stay in Louisville just for the winter while he worked up money to rent a place in Santa Rosa, but the rent prices in the city remained high all the way through the spring, and he didn't want to move the family back until he was sure he could keep up with a new lease. At the end of six months, he told his family they would have to rent in Louisville instead, where housing was cheaper and where they had far more room than they would have in Santa Rosa. The children hated the weather in Kentucky, and José didn't like it much, either, but it felt like the door had shut behind them when they drove out of California.

The longer the family stayed in Kentucky, the less certain José became that they would ever go back. The insurance policy wasn't enough to re-build the old house, but he still owed around half the mortgage, and he wouldn't be able to break even just by selling the burned-out lot. As the months went by in Louisville, José went back and forth about what to do— sell the lot and cut ties with Santa Rosa, or drain his savings to rebuild?

As he mulled over the decision, the housing market in Santa Rosa con-tinued to deteriorate. Two years after the Tubbs Fire, just as the first victims were returning to their old homes, another blaze called the Kincade Fire raged down from the northern mountains and licked its way toward Santa Rosa. This time firefighters controlled the blaze before it reached the city itself, but the fire still took out close to two hundred homes in the suburbs of Healdsburg and Geyserville, putting additional pressure on the hous-ing market. Around the same time, a large homeless encampment emerged on a bike trail in central Santa Rosa, becoming a sanctuary where people who had been unhoused for years slept alongside those who had lost their homes during the scramble that followed the fires. When the coronavirus pandemic began the following spring, and white-collar remote workers beat a retreat from San Francisco, rents in the area surged even higher, push-ing more low-income families out of the housing market. A few months later, yet another fire gnawed away at the region's already depleted housing stock: the Glass Fire shot the gap between the burn scars of two previous blazes, destroying more than six hundred homes in a section of the hills that had been spared by the Tubbs event. Rents reached new highs, as did the county's homeless census. The affected neighborhoods in 2020 were high-income suburbs like Fountaingrove, and some of the new victims had lost their first or second homes in the 2017 fires as well—they had moved, but not far enough to escape the risk. These newly displaced now searched wherever they could for temporary accommodations.

Wildfire had gone from a once-in-a-lifetime disaster to a fact of life for almost everyone who lived in California. Even if your home didn't burn, there was always the looming threat of an evacuation, or the torture of spending weeks beneath a pall of choking smoke. Wealthier people in the area started to leave town for second homes during the summer and autumn, becoming "firebirds" who reversed the typical snowbird pattern; those who couldn't afford to leave stayed inside during smoke warnings and gave up on their morning walks. The Tubbs Fire was no longer just "the fire," it was one in an endless series of assaults on what had once seemed a safe place to live.

It took until the winter of 2021 for José and his wife to decide what to do about their lot in Coffey Park. They would take their meager insurance payout, rebuild a smaller house on their old lot, and hope California would be kinder to them the second time around. Neither José nor his wife had ever liked Louisville, but even so their decision was only made possible by the life changes that followed the fire: the oldest Guzman son had moved out to Los Angeles, and the next oldest was in his midtwenties and would soon leave the nest as well. They couldn't afford to build the same house they'd once had, but they no longer needed it.

José felt relieved to be back in Coffey Park, but he and his wife still weren't sure they would stay in the neighborhood for the long term. The firm hand of the market and the ever-present risk of fires had already started to nudge them out again. The home values had risen so high that José now thought he could sell the rebuilt house for six or seven hundred thousand dollars, enough to pay off the remainder of his mortgage and start fresh in a more modest house somewhere else with lower risk. Maybe then, more than half a decade after Tubbs, he and his family would be able to say they had recovered.

Still, there was always the question of where the next fire would hit—was there any part of Sonoma County that was guaranteed not to burn? Just like in Big Pine Key after Irma, the aftermath of the Tubbs event brought a realization as dark and as suffocating as smoke: the crisis was not going away.

III.

Tipping the Scales

In the summer of 1823, a young priest named Father José Altimira set out with a small group of soldiers to explore the hill country north of San Francisco. Altimira was a member of a religious order that had established a group of Catholic missions in and around San Francisco, and his superiors had dispatched him to survey the area to the north for suitable building sites.

As Altimira rode through the Napa and Sonoma Valleys, he noticed something strange in the countryside around him. There were large burn scars along all the hills, huge sections of charred land that the Indigenous tribes seemed to have burned only a few weeks earlier. "[The hills] were soon to be burned of the long grass by the Indians we met," he wrote in his diary. "The place is bare of thick woods."

Fire is an essential stage in the life cycle of forests like the ones in Northern California. Over the course of decades, scrub plants like chaparral accumulate on the forest floor, creating a dense undergrowth amid tall conifers; periodic fires help clear away these tangles of old growth and make room for new grasses and flowers. For as long as the Miwok and Wappo peoples had lived in the mountains they had engaged in careful stewardship of this forest environment. Families and small communities were responsible for carrying out controlled burns in the areas where they lived, ensuring that vegetation on this forest floor did not pile up. There was no central political entity that supervised the overall forest environment, but the forests remained healthy nonetheless.

When American settlers arrived in the 1800s and drove out the Indigenous population, they also attempted to impose their own firefighting practices on the ancient forests. To protect the all-important timber industry, the federal government pursued an aggressive program of fire suppression and prevention. The US Forest Service instituted the so-called 10 a.m. policy, which decreed that all fires had to be extinguished by ten o'clock the morning after they began, and hired thousands of firefighters and smoke jumpers to enforce it. The state of California had its own firefighter troop, too, made up in large part of inmates from state prisons. The federal government also embarked on a massive education campaign that urged people to prevent fires at all costs, anchored by the now-iconic character of Smokey Bear. Subsequent scientific research confirmed what the Miwok and Wappo had always known—that the forests needed to burn in order to thrive—but the federal government's policy never caught up with that research. The protection of built-out cities like Santa Rosa, meanwhile, fell to the state agency, Cal Fire, which followed the Forest Service's lead and adopted a policy of total suppression.

Even so, it was impossible for the government to suppress *all* the fires, especially in the areas where human habitation overlapped with wild forest. There had been five major fire events in the hills around Santa Rosa since the mid-1800s, culminating in the massive Hanly Fire of 1964, which blazed the same path that the Tubbs Fire would trace over half a century later. But in contrast to Tubbs, the Hanly Fire destroyed a mere one hundred homes: at the time the hill country was still largely undeveloped, aside from a few vineyards and farms.

In the aftermath of the Hanly event, Santa Rosa city leaders deemed the Fountaingrove hills too vulnerable to support large-scale residential development. If a few individual homeowners wanted to live in the fire zone, the city wouldn't stop them, but they weren't going to support the construction of whole new subdivisions and neighborhoods. As the decades wore on, however, the memory of the fire grew dimmer, and the city reversed its decision. Napa and Sonoma saw a surge of growth from winemaking and luxury tourism, and Santa Rosa was selected as the site of a new corporate headquarters for the computer giant Hewlett-Packard. A lot of wealthy people wanted to live in the area, and if Santa Rosa didn't try to capture their tax dollars, they would build homes elsewhere. The federal government had also greased the wheels by offering Santa Rosa new money to build a parkway that would arc through the hills.

Faced with all these incentives to build, the city chose growth over safety. It approved the construction of several luxe subdivisions that looped up and down the hills above Santa Rosa, allowing developers to plant hundreds of massive mansions in thick pine forests and on sheer rock cliffs.

The glue that held this whole venture together, that legitimized all this construction in risky areas, was the private insurance market. The average homeowner's insurance policy covers a range of potential hazards such as burglary, hail, and lightning strikes. Flood damage is not included in traditional coverage because private insurers cannot turn a profit covering it—major insurers had stopped selling flood policies in the 1920s after high claims almost drove them out of business, leading the federal government to step in. Fire, on the other hand, was still profitable for the insurers that sold policies in California, in part because wildfires usually weren't large enough to threaten residential property. There were brush blazes in dry years, and isolated events in the hilliest areas, but for the most part the Forest Service and Cal Fire had always managed to snuff out these conflagrations before they reached the scale of something like Tubbs. The rapid development around Santa Rosa took place during a fifty-year reprieve between major fires, a period when the insurance companies that served the area had never had to make large claim payouts. For as long as there were no fires in Santa Rosa, there was no reason for insurers not to offer fire coverage to the people who lived there, and if those people had insurance, there was no reason for banks and lenders not to offer them mortgage loans. The insurance market socialized the risk of fire damage across the state, creating an equilibrium that allowed people to live in areas that were destined to burn.

It was climate change that tipped the scales and upset this delicate balance. A profound drought had racked the West in the years before the Tubbs Fire, drying out huge swaths of vegetation and priming California to ignite. The climate-enhanced dry spell combined with warm autumn temperatures to create the ideal conditions for fire, not just in Santa Rosa but across the entirety of the state. The tacit assumption of the insurance system had always been that federal and state firefighters could keep major blazes under control, but now the fires burned so hot and so fast that nobody could contain them. The climate crisis might not have been the proximate cause of the Tubbs Fire, but it made such events far more likely. What had once been unfathomable was becoming normal.

The Tubbs fire revealed that the insurance equilibrium in the state relied on flawed math. The home prices in a housing-scarce city like Santa Rosa could get up into the millions, but most fire policies provided only a few hundred thousand dollars' worth of coverage, insuring only a portion of a structure's value. This made sense to an extent, since electrical fires and other small blazes might only damage part of a home rather than level it altogether, but it was also a financial decision on the part of the insurance companies. The companies knew that wildfires were still possible, and they didn't want to be liable for rebuilding an entire neighborhood's worth of mansions, so they offered only enough coverage to satisfy lenders and homeowners. Thus, in the aftermath of the Tubbs Fire, many victims like José Guzman found that their insurance policies were not sufficient to cover the cost of rebuilding the homes they had lost, and one's insurance policy set a ceiling on the value of one's future home. It didn't matter if a victim rebuilt their home on the same lot or bought a new home somewhere else. Unless she had extra savings to make up the difference, she could only get what her insurance payout would buy.

Strange as it might seem, this dynamic gave the middle-class home-owners in Coffey Park a distinct advantage over the wealthy homeowners in Fountaingrove when it came time to rebuild. Because the houses in Coffey Park were more modest than those in Fountaingrove, there was a smaller gap between the average insurance policy and the average home value, so most victims could afford to build new houses that were similar to the ones they had lost. The sheer scale of destruction in Coffey Park also proved to be a blessing in disguise since it attracted several major home-building companies to the area. These home builders offered neighborhood victims a bulk dis-count if they picked from a list of predesigned, cookie-cutter houses, some of which were even larger than the original models. It would take years for the neighborhood to look like itself again, and in the meantime the dislocated families would have to scramble for housing, but the relative affordability of the area ensured that the fire-driven exodus would only be temporary.

Nevertheless, Kevin Tran's parents weren't sure at first that they wanted to rebuild in the neighborhood. The loss of all their photo albums and keep-sakes had weighed on Kevin's mother, and his father worried that another fire would strike in future years. Kevin, too, had not forgotten the sight of trees bursting into flame as he and his family fled the neighborhood in the dead of night. It was a difficult decision, and the Trans didn't have long to make up their minds: their insurance company would only pay for two years of temporary living expenses, and it would take at least a year to clear the burned-out lot and build a new house.

We tend to think of moving decisions as personal decisions above all else—a family weighs its love for a home against its fear of another fire— and it's true that the Trans had personal reasons for deciding to stay. Kevin's mother and father had been planning to close the restaurant and retire even before the fire, and they wanted to be somewhere familiar. Even so, the primary factors in the decision were economic. The Trans might have considered moving to a similar neighborhood with lower fire risk, and their insurance company would have been more than happy to see them use their claim payout to go somewhere safer, but there was no such neighborhood in Santa Rosa or indeed in all of Wine Country. The housing crisis had squeezed the area's middle class almost to the point of nonexistence, and for the Trans, moving elsewhere would have meant upgrading to a much more expensive property. Their insurance policy gave them only enough money to build a Coffey Park–caliber house, and there was only one Coffey Park. Moving back was a difficult decision, but it was also, in many respects, an easy one. Most other residents I interviewed had the same ambivalent journey as the Trans did: they liked the neighborhood, and they liked the idea of moving back, but they also knew they didn't have much of a choice. Unless a victim wanted to downsize or move far afield, they had a significant financial incentive to stay put.

The Trans were one of the first families to move back to Coffey Park, and as they settled into their new home, they watched a spiffier and more modern version of the neighborhood they knew bloom into existence around them. The new houses were larger and better built than their 1970s-era predecessors, with trimmer lines and a cooler color palette. These new structures lacked the idiosyncratic charm of the old neighborhood, but many returning residents now had an extra bedroom or even a second floor. The neighborhood's central park got a new playground and a set of ping-pong tables, a welcome-home investment from the city of Santa Rosa and several nonprofits. The fire had taken all the cars in the neighborhood, too, except for Kevin's sprinkler-protected Verizon sedan, and many residents had used their payouts to buy upgrades and new-model editions—one returning homeowner even splurged on a vanity plate that read THXTUBS. For the Trans, this resurrected neighborhood felt like a safer and more pleasant place to live than the neighborhood they had watched go up in smoke. As the months went on, they found that almost everyone else in the neighborhood had also opted to return, since Coffey Park was the one place where their claim payouts would allow them to rebuild. Kevin's sister returned to the neighborhood, building a new home on her old lot right around the corner. Both the neighbors on either side of him also returned, as did the family across the street. The group of friends with whom he had played *Pokémon Go* also came back, except for one whose parents had been renters and who ended up moving a few towns over. The neighborhood looked different and felt different, but when Kevin saw the same familiar faces walking dogs and playing in the park, he felt like he was home.

—

Things did not go so smoothly in the hills of Fountaingrove, where the same insurance dynamics created a crisis that foreshadowed broader turmoil in the California property market.

Vicki and Mark Carrino got an insurance payout, too, and they used it to rebuild the home they had escaped on the night of the fire. Like many disaster survivors, their attachment to their property had only grown stronger, and they wanted to build as close a replica of their old home as they could. Their insurance policy wasn't large enough by itself to cover the cost of a new house, but the income from Mark's construction business made up the difference.

Nevertheless, rebuilding the house in Hidden Hills was far more difficult than rebuilding a bungalow in Coffey Park. The neighborhood's position atop the Mayacamas foothills had once signified its residents' position atop the city's economic ladder, but now the high elevation felt more like a curse than a blessing. The Carrinos had to haul construction materials all the way up their spiral driveway, hire an army of engineers to help ensure it was safe to build on their steep-sloping lot, and negotiate for months with the city and the county before they could secure the permits they needed for such a complex project. The last and most difficult task was the assembly of a cantilevered swimming pool where their back deck had been—with no trees left to give them shade from the sun, the couple figured they needed a way to cool off.

Building a house is a favorite prerogative of many wealthy couples, and for the Carrinos the post-fire construction project was a rewarding outlet, a difficult but worthwhile use of time and money. For the older couples that lived in the hills around them, though, rebuilding in the neighborhood was inconceivable. Most couples did not have enough money to make up for their insufficient insurance policies, and many didn't have the energy to deal with years of contractor delays and construction headaches. The Carrinos made it back to Fountaingrove two years after the fire, but their neighbors never followed in their wake. The man who owned the home to the north sold his house to a winery, and the couple to the south rebuilt only to sell, as did two other couples on the Carrinos' street. The street that ran parallel to theirs on the next ridge turned over in its entirety, with every former resident passing their lot to a new owner, and many of the lots remained empty even years after the Carrinos moved back. The subdivision had always been secluded, but now it felt downright lonely, even almost abandoned. Before the fire, Vicki could look out from the back deck and glimpse her neighbors' houses in the spaces between the trees; now her view was wide open, unobstructed, but there were almost no houses on any side, just empty yards and the shells of old propane tanks. Farther off, on more distant hills, she could see new structures rising at long last, but she had never met the people who lived in them, and she wasn't sure she ever would.

Even half a decade after the fire, Fountaingrove still looks far from whole. Indeed, to visit the neighborhood today, one would think the recovery had just begun. The insurance gap in the area was so extreme that many residents chose to cut their losses and buy somewhere else rather than spend countless thousands of dollars to rebuild homes in an area they knew was dangerous. They sold their empty lots to investors and speculators from out of state, most of whom sat on their holdings and waited for the price of land to rise rather than build new houses. The "neighborhood" that remains in the wake of this insurance-driven exodus is a husk of its former self, a community with gaping holes in it, looking almost as if it had never been finished in the first place. You can sense the emptiness from the moment you begin to ascend Fountaingrove Parkway, passing through a complex of office parks and into the remnants of old subdivisions. Most of the empty lots are still cluttered with piles of debris and burned-out tree branches, and the minority that have been cleaned up now bear aggressive FOR SALE signs. Even the public infrastructure in the neighborhood still bears quite a few bruises from the fire: there are weeds sprouting in medians and planters all along the parkway, and the charred shopping mall sign at the entrance to the neighborhood has never been torn down. If the insurance market ensured that displacement from Coffey Park was temporary, it has also ensured that displacement from Fountaingrove was permanent.

In one sense, this was how insurance was supposed to work. The families in Coffey Park had purchased enough coverage, so they got their homes back; the families in Fountaingrove had not, so they did not. Every customer received a payout proportional to the amount of risk they had agreed to cover. In another sense, it was shocking: insurers and their customers had underestimated the potential for climate-driven fire damage, and now the customers were being asked to shoulder the massive financial burden of rebuilding.

The insurance companies suffered, too: they had enough money to pay out any individual customer in Coffey Park, but they weren't prepared to pay them all out at once, and the wholesale reconstruction of the neighborhood dealt a gut shot to the companies' annual profits. To make matters worse, the insurers didn't have any control over fire-safety standards for the rebuilt homes, so the new version of Coffey Park turned out to be just as vulnerable to wildfires as the old one. Since the neighborhood was outside the wildland-urban interface, protected by the freeway, it was exempt from the stricter building codes that the city of Santa Rosa adopted after the fire. The new homes were still spaced just a few feet apart and separated by wooden fences, and many of the new yards were planted with flammable mulch. The new home builders also declined to pay for fire-prevention upgrades like hardened concrete walls and fireproof windows. It might be a long time before another fire skips over the freeway, but when it did, the insurance companies would again get stuck with a massive bill.

And the worst fires were yet to come. Just a year after the Tubbs event in Wine Country, the Camp Fire obliterated the town of Paradise on the eastern edge of California, incinerating more than one hundred thousand acres and destroying thousands of structures. The insurance companies that covered Santa Rosa were the same ones that covered Paradise, and now these companies had to pay out billions more dollars to the victims of the latter fire. The combined insurance claims from the 2017 and 2018 fire seasons surpassed $12 billion, wiping out twenty-five years of cumulative underwriting profits, and the next three years were little better.

The wildfires had already created an ecological crisis in California, but now they were also stoking a financial crisis in the state's housing market. The self-restoring forest that predated colonization had turned into an enormous tinderbox, and millions of people were living in areas primed to burn. The total risk burden in the state had grown so large that insurance companies could no longer spread it out among their customer bases—in other words, they couldn't maintain profits if they kept having to pay out so many victims every year. These insurers soon reached the same conclusion that FEMA had reached after Hurricane Floyd: they had to reduce their financial exposure.

On the second anniversary of the fire, Fountaingrove residents Lisa and Damon Mattson received a letter from their insurance company, Homesite, a partner of Progressive. The company informed the couple that their policy had come up for renewal and that Homesite was planning to quadruple their premiums, raising them to $8,000 a year.

The Mattsons are the image of Fountaingrove wealth. Lisa is the creative director for Jordan Vineyard & Winery, a high-end concern best known for its Cabernet Sauvignons, while Damon has held senior positions at a few medical technology companies. The Mattsons' home survived the initial onslaught of the Tubbs Fire, but it burned a few days later when a stray ember landed on a Crate & Barrel lawn chair and sparked a fire that destroyed almost the entire structure. The blaze continued down the hill and knocked out more than a hundred trees, denuding the ridge and exposing a view of the charred slopes.

The damage to the Mattsons' home exceeded one and a half million dollars, but their insurance policy only provided about half a million dollars of coverage. Here, too, the gap was intentional: the couple's insurance agent had recommended they purchase a low level of coverage, and Damon had received the same result when he used the company's online calculator. The insurance company had weighed the total value of the house and its contents against the risk of the fire and generated a number that represented a compromise between the two, one that wouldn't cover the cost of rebuilding.

It cost the Mattsons more than a million dollars to restore their home, and the intangible costs to their well-being were even greater—the couple took out a second mortgage, wrangled with dozens of unreliable contractors, and suffered through a contractor-botched revamp of their windows and floors. Halfway through the process, Lisa suffered a nervous breakdown and confined herself to bed rest for more than a month. At last, two years after the fire, they returned to the neighborhood along with a few of their luckier neighbors.

The new home was designed with an eye toward fire protection: they replaced the grassy backyard ridge with a desert-themed array of drought-tolerant plants and installed a hurricane-strength wind protector to guard against the powerful Diablo gusts. Damon even bought a full firefighter suit that he planned to don if there was ever another fire in the area. The house wouldn't stand a chance against another Tubbs Fire, he admitted, but in the event of a smaller blaze he felt confident he could stay and defend the house against rogue embers. They also negotiated an upgrade to their insurance plan, expanding their coverage to the home's full value.

Despite all that, Homesite raised their premiums almost as soon as they moved in. The company had paid out billions of dollars of fire claims in California over the previous few years, and it couldn't afford to have home-owners like the Mattsons on its balance sheet anymore unless they were willing to pay an exorbitant cost to cover their risk; the Mattsons thought their new annual premium was outrageous, but Homesite was saying it was outrageous for them to live where they did.

Lisa and Damon scrambled to find an alternate insurer, but every other company that served Santa Rosa either offered them a similar quote or told them they wouldn't take new customers in Fountaingrove anymore. In the end, the couple gave up and enrolled in the California FAIR Plan, a state-administered public option for people who can't obtain insurance on the private market. The FAIR policy was less expensive than the out-rageous new Homesite policy, but it was still far more than the couple had paid for insurance in the past, and the coverage capped out at a few hun-dred thousand dollars, far less than what the Mattsons' home was worth. Four years had passed since the fire, but the Mattsons were worse off than they started, with an expensive insurance plan that did not cover the value of their home.

The Mattsons weren't the only family to receive such a letter. Home-owners across Wine Country saw their premiums spike to unreasonable heights after Tubbs, and new arrivals to the area were hearing that the com-panies had decided to stop issuing new policies. Elsewhere in the state, in regions that had *not* burned, the situation was even worse: insurers had started to cancel policies for existing customers, dropping thousands of clients from their rolls to decrease their exposure to wildfire risk. The state of California prohibited companies from dropping customers in areas that had experienced wildfires in the past year, but this policy only created a game of whack-a-mole in the state's insurance market, leading companies to drop customers who had never filed a claim or so much as witnessed a wildfire. If a fire burned in the Sierra Nevada one year, the companies dropped customers in areas like San Mateo, which had been lucky enough to go unscathed; if San Mateo burned the next year, the companies dropped customers in the Central Valley. Local newspapers were filled with stories of customers in Oakland or Sacramento or San Diego whose insurers dropped them for no clear reason. The wineries that sustained Sonoma County's economy also saw their insurers raise their premiums or drop them altogether, which left their multimillion-dollar operations unprotected heading into another fire season. Over all these households and businesses loomed the threat that mortgage lenders might find out they didn't have coverage and swoop in to demand collateral. The state government scrambled to find a solution, but there was no easy fix: the homes had already been built, and people were already living in them, so in many respects it was too late to intervene. The government couldn't force companies to lose billions of dollars selling unprofitable insurance, but if officials allowed companies to pull out of the California market, they risked popping an enormous asset bubble. The state and the insurance industry entered a kind of holding pattern, each party waiting for the other to make the first move.

Santa Rosa ended up seeing no major fire activity in 2021. The Dixie and Caldor Fires were more than a hundred miles to the east, in the distant mountains of the Sierra Nevada, and the endless Windy Fire in the Sequoia National Forest was hundreds of miles to the south. This should have been a relief for the residents of Fountaingrove, but it was hard to feel reassured. If there were no fires in the area, there would be no insurance moratorium, and if there was no moratorium, there would be nothing to stop insurers from dropping customers all over the city. In the new climate-fueled calculus of fire risk, even the spared were not spared. To live in the state at all was to accept one's own share of the financial burden of climate change, a burden that with every year and every new fire grew harder to bear.

I visited the Mattsons toward the end of the 2021 fire season, just weeks before a heavy rainstorm brought some much-needed moisture to the parched countryside. As we sat out on the back deck of their new house and took in a breathtaking view of the hills, I could see that the forests around Fountaingrove had not even begun to grow back. The slopes were still barren for miles in every direction, so Lisa could point out the separate burn scars from the Tubbs and Glass Fires on the mountains in the distance. This lack of new growth meant that there was very little flammable material around the Mattsons' house, but there was a downside as well: the autumn wind now made a ferocious noise when it scraped across the barren hill. Another couple in the Mattsons' subdivision had started to rebuild on their old property but had decided to leave after they heard the screech for the first time.

Despite all the time and money the Mattsons had poured into restoring their house, they, too, were uncertain whether they'd stay up on the hills. Having gone deep into debt to finance the rebuild, the couple had pinned their hopes on a potential settlement from Pacific Gas and Electric, the electrical utility whose infrastructure had caused several of the 2017 wildfires. The utility had promised to pay out billions to Tubbs victims, and the couple was hoping that the settlement would be enough to help them break even. If not, they would put the house on the market and move somewhere else—perhaps down the hill, they said, or perhaps into the valley, or perhaps out of California.

For the moment, though, there was nothing they could do except sit on the deck and listen to the wind.

CHAPTER FOUR

The Story of the Verdins

COASTAL EROSION AND

CULTURAL EXTINCTION

Pointe-au-Chien, Louisiana

I.

Eden Interrupted

T he Mississippi River is the grandest and mightiest natural force in North America, pushing more than three hundred billion gallons of water into the Gulf of Mexico every day. The river has no single point of origin: thousands of tributaries twist together into its berth the way nerves twist around a spine, draining water from more than half the landmass of the continental United States. Suspended in this water is sediment, thousands of tons of it, the castaway dirt and dust of more than thirty different states. As the river enters the gulf, it lets this sediment go, building forks and fans of fertile land. The soil accretes so fast that the ocean cannot erase it.

The coastline of Louisiana is the tentative result of a battle between the Mississippi River and the Gulf of Mexico, a battle that has lasted thousands of years. The river is always switching direction as it seeks the path of least resistance, and the ocean is always fighting it, eating away at the sedimentary land. This is why the coastal part of the state is not a series of beaches but a maze of inlets and islands and lakes. This watery maze is called "the bayou," and the individual streams in the maze are also called "bayous," as though each small waterway contained the complexity of the swampy whole. For most of its history, the bayou was a bountiful place, protected and profuse with life.

For as long as the stalemate between river and ocean held firm, isolated and self-sufficient communities thrived in the bayou, developing their own cultures and even their own languages. But over the past half century, as industrial development and climate-driven sea-level rise have eaten away at the bayou land, these communities have faced first exodus and then extinction. This exodus was not the result of a single disaster, nor was it just a matter of people losing their homes—the rising water altered the facts of life on the bayou over the course of generations, reversing the relationship between a people and their land and leading many to move inland of their own volition. Those who stayed behind saw the places they knew slip away as islands vanished, jobs were eliminated, and traditions were lost. The collapse of these bayou communities reveals a profound truth about what we stand to lose in the coming era of climate migration—not just homes and subdivisions, but entire cultures.

We don't know exactly when the first people came to Bayou Pointe-au-Chien, a slow-moving waterway about forty miles south of New Orleans. The evidence suggests that a succession of Indigenous tribes must have migrated to the area over the course of several centuries, perhaps as offshoots of dominant mainland tribes such as the Chitimacha. These large tribal organizations had maintained a presence in and around the Mississippi and Atchafalaya river basins for hundreds of years, but it was not until the eighteenth century that they began to make a permanent movement south. As the Choctaws of Mississippi made more frequent contact with English colonists, some bands migrated west to find an undisturbed refuge; at the same time, the Chitimacha fled south and west to escape the murderous offensive of the French colonial administrator Jean-Baptiste Le Moyne de Bienville. They joined with the remnants of a smaller tribe called the Houma, and these three Indigenous bands staked a claim to the dense and uncharted marshland of what is now Terrebonne Parish—the name means "good earth," but it's a misnomer, since the parish area is more than two-thirds water. The French colonial government in the port city of New Orleans did not maintain a permanent presence that far south in the marsh, and, at first, the migrating tribes had the land to themselves.

Soon after the Chitimacha and Choctaw established permanent settlements along the bayou, though, another group of displaced people arrived in New Orleans. In 1755, the British army seized the French colony of Acadia, located in what is now eastern Canada; the Redcoats rounded up and deported more than ten thousand French settlers in a violent ordeal known as Le Grand Dérangement. Many of these scattered Acadians ended up in the fast-growing city of New Orleans, where they and their fellow Frenchmen came to be known as "Cajuns."

One of these settlers was a man named Alexander Verdin (sometimes spelled Verdun or Vardin), born in New Orleans around 1770. At age thirty, Alexander married a woman named Marie Gregoire, identified in contemporary records as a *femme sauvage*, or an Indigenous woman. Interracial marriage soon became illegal under Louisiana law, and the seven children whom Verdin's will calls "free colored men and women" were all considered illegitimate by the state. Land transfer records indicate that the couple obtained a few plots of land from other "men of color," and by the time Verdin died in 1833 the family had established itself in the area around Bayou Pointe-au-Chien.

There were several other such interracial marriages during that period, all of them between French colonists and Indigenous women, and within just two generations this handful of marriages had produced an expansive family lineage. The descendants of Alexander Verdin spread out down the length of the bayou and journeyed east and west to other sections of the marsh, intermingling with other farming and fishing villages along the coast. After another two generations, the family lineage of the bayou had blossomed into an independent and informal tribe, a subculture whose members spoke an accented French and whose religion incorporated elements of Indigenous ritual and Catholic sacrament. Most of the people who made up this tribe could trace their lineage back to the same few intermarried families, and that is still the case today: a federal government report published in 1994 found that there were four thousand tribe members who used just four common surnames.

For more than a century the Verdins and their descendants enjoyed almost complete freedom from the mainland. They established outposts all the way down the serpentine length of Bayou Pointe-au-Chien, gathering together in towns and villages. The waterway was narrow and shallow, canopied by ancient oak trees, and a newcomer to the area would struggle to navigate through the labyrinthine marsh and down to the lower reaches of the bayou. They lived in houses built out of cypress wood and thatched with palmetto leaves, and they navigated up and down the narrow bayou with cypress rafts and dugout canoes.

Despite the bayou's proximity to the Gulf of Mexico, and despite its position just a few inches above sea level, the early residents of the area didn't have to worry about flooding, thanks to the thick marsh forests that surrounded them. Their palmetto houses did not sit on stilts above the ground as houses do today, and they did not need levees or floodgates to protect their villages from storms. In the rare event of a hurricane strong enough to overwhelm the marshland barrier, local legend says that the members of the tribe would stand in a circle around one of the burial mounds and tie themselves with a rope so that they'd stay anchored together even if the water washed up over the islands.

In addition to protecting them from storms and tides, the fertile soil of the bayou provided the Verdins with everything they needed to live off the land. They raised chickens, horses, and cattle on the solid soil, letting the animals roam back and forth across the bayou. The marshland barrier also protected the fertile island soil, making it possible for every resident to raise his or her own family garden and to weave baskets out of palmetto fronds. The range of fruits and vegetables that could thrive in the delta was astounding: tribal elders told one researcher that in the old days they had been able to grow "butter beans, green beans, lima beans, potatoes, cantaloupe, watermelon, okra, cucumbers, peas, mustard greens, carrots, corn, and rice."

The healthy marshland that once encircled the bayou was also home to thousands of muskrats, minks, and nutrias, a rodent that grows to about the size of a raccoon. These rodents were perhaps the greatest blessing of all, since they allowed the tribespeople to make money from fur trapping, which was still an essential part of the French colonial economy. On a good day the average family could collect and treat as many as twenty-five pelts, and the pelts would sell for as high as three dollars apiece, forty-five dollars in today's money. The tribespeople also had a fruitful relationship with the water that surrounded them: the thousands of freshwater islands out in the marsh provided a perfect place for shrimp to lay their eggs, so that even a child could wade out into the swamp and come back with a full net. The same was true for oysters, which thrive in the estuarine water of the bayou. If for some reason a family needed extra money, they could always row their pirogues up the bayou to the nearby town, where French and English settlers had established large sugarcane plantations. There was some mutual suspicion between the white and Indigenous sections of the bayou, but for the most part the two sections left each other alone, and they were happy to do business with each other.

For the better part of a century, the nascent tribespeople of Pointe-au-Chien remained undisturbed in this marshland Eden—they asked for little from the mainland, and in return the mainland left them alone. Over the course of five, six, seven generations they had created an almost utopian community, a society untouched by the forward motion of time.

Like all eras of peace, though, this one could only last for so long.

—

In the late 1920s, on the eve of the Great Depression, a newcomer arrived in Louisiana, wearing a three-piece suit and sporting a thick mustache. His name was Joseph S. Cullinan and he was the founder and chief executive of the Texas Company, better known today as Texaco. A native of Pennsylvania, Cullinan courted northern investors with promises of an independent outfit that could compete against monopolistic giants like Standard Oil. He had started with a small drilling operation along the Gulf Coast of Texas and expanded in both directions over the years, moving west toward Houston and later pushing east toward New Orleans.

The Verdins didn't know it at the time, but their village sat on the border of prime subterranean real estate. The sedimentary outflow from the Mississippi River had over the course of the previous millennia created large domes of salt on the ocean floor just a few miles away from the bayou, and large deposits of oil and natural gas had accumulated beneath those domes. Cullinan and his fellow oil tycoons wanted to drill underwater, pump the oil up to the surface, and refine it into gasoline.

In every sense but one, the land around the bayou belonged to the tribe—they lived on it, traveled on it, shrimped on it, raised cattle on it, trapped on it, and grew their gardens on it. They knew its every intimate detail. From a legal standpoint, though, the land was the property of the Delaware-Louisiana Fur Trapping Company, a land-speculating outfit that leased sections of marshland to amateur trappers from the mainland. The company owned almost all of the marshland in the parish, and it soon started suing bayou residents who went trapping without a lease—a court judgment from November 1923 enjoined the Verdins from trespassing on the company's trapping areas.

When oil barons like Cullinan arrived on the Gulf Coast, the company got serious. It rebranded as the Louisiana Land and Exploration Company and started leasing out larger sections of land at much higher prices, acting as a land broker for oil companies like Texaco. More oil companies like Exxon, Mobil, and Conoco arrived over the next two decades, leasing thousands of acres of land and constructing dozens of drill derricks and pipelines.

The oil companies started in the distant sections of the marsh, close to the barrier islands that separated the bayou from the open gulf, but over time they started to encroach on land where the residents of the bayou were accustomed to raising cattle and cultivating beds of oysters. An elder who lived on an island near Bayou Pointe-au-Chien later recalled that "every year or two years they would get closer and closer until finally we had them on our fence posts." The settlers who lived farthest down the island could hear the rumble of the drills from their beds and see the derricks from their yards.

In 1933, Louisiana Land renewed its legal attack against the entrenched bayou residents, suing dozens of tribespeople including Lawrence, Pascal, and Octave Verdin, all great-grandsons of Alexander Verdin. The trespassing suit was filed in federal court in New Orleans, almost fifty miles away from the bayou, which meant it was almost impossible for the defendants to make an appearance in court and plead their case. Even if they had been able to show up, the tribespeople might not have been able to understand the proceedings, since their native language was a dialect of French and not the English of the courtroom. In the absence of their input the federal judge "ordered, adjudged and decreed that a permanent injunction be issued . . . strictly commanding and enjoining" the defendants to keep off the company's land.

Within a few decades Louisiana Land had all but triumphed over the residents of Pointe-au-Chien: the tribespeople were no longer free to trap and fish wherever they wanted, and those who lived farthest down the bayou had to move north onto safer land to make room for the armada of new oil and gas wells. But the industry did not stop there. Ferrying the oil from the marsh to the mainland demanded the construction of myriad canals and shipping channels. Louisiana Land sent huge dredging vessels back and forth through the swamp, clearing new paths for the water in what had been a dense estuarine environment and shredding the delicate labyrinth of the marsh. The boats made a horrendous roar as they crashed through the deep-rooted cypress trees and dredged the dirt up from underneath the water. The uninterrupted sweep of the marsh came to look like a web of abandoned streets and alleys, the grid plan for a watery ghost town. When the oil giants transitioned to offshore drilling on massive rigs in the outer gulf, the state and the federal government cofinanced the excavation of two massive navigation canals, comparable in width to the Mississippi River.

As the oil companies sliced up the bayou to make room for their ships and pipelines, carving dozens of canals and sanding islands out of existence, they removed the barriers that had been instrumental for centuries in holding the water back. The extraction of oil from the subterranean salt domes also reactivated subsurface faults along the bayou, which in turn caused masses of marshland to slide down into the water. To make matters worse, the Army Corps of Engineers had built hundreds of levees along the Mississippi River to prevent farms from flooding, which had halted the inflow of new sediment to the delta. There was no new land being built, and less land than ever to stop the water from rising. The solid marshland soil grew moist, then water-logged, then disappeared beneath the high tides. The oak and cypress trees withered and ceded ground to saltwater grasses. Each new storm and heavy rain wiped away another island, carved a peninsula in half, smeared strands of water across the mudflats. The floodwaters pushed up the narrow reach of the bayou itself, widening the waterway like a drill pushing a screw into a wall.

This process of erosion took place over the course of decades, so slowly in fact that even the people who witnessed it have a hard time describing what it felt like. If you ask elderly residents to describe the change, they often do so by telling a story, describing a binary change between before and after. One man who grew up on the bayou remembers when he and his brothers could play games on land of which not the barest trace remains. Another recalls a time he set a pinto pony loose and it later appeared three bayous over, having plodded across solid ground that is now open water. Another woman recounts a childhood spent rowing down a stream near her house and touching the grass on both sides; now the stream has grown so wide it can accommodate a tugboat. The water subsumed graveyards, subsumed the bottoms of burial mounds, subsumed the old sites of towns that had names and histories—Fala, L'Esquine, Bayou de la Valle, and more. Over the past century, Louisiana has lost around two thousand square miles of land, an area larger than the state of Delaware.

The worst, of course, was yet to come. It would be another few decades before the human-caused warming of the oceans increased and water levels on coastlines around the world began to rise at unprecedented rates. Nevertheless, there is a devilish irony in the fact that it was the fossil fuel industry that first destabilized the land around the bayou. The companies dredged canals in marshland and drilled holes in the ground so that they could extract oil and gas from the earth. They sold that oil and gas to people who burned it to fuel their stoves and cars. The combustion of those fuels released carbon dioxide into the atmosphere, and the oceans absorbed much of that gas, which caused them to warm up and expand; the unleashed gas also raised the earth's surface temperature, which caused glaciers to melt, which caused the ocean to expand further. As the ocean expanded, it subsumed more sections of coastal land, but that land had already been sinking as the oil companies vivisected the marsh.

In other words, the oil companies submerged the bayou so that they could make money from a product that would later submerge the bayou even farther, forcing families like the Verdins to leave their homes and move inland. Like a prophecy drawn on the wall of a cave, the initial action foreshadowed the ultimate result.

II.

Oil and Water

F or Wallace Verdin, 1964 was a year of many changes. He moved into a new house, he got a new job, and his children moved to a new school. This sort of thing happens to many families, but for Wallace it was different—it felt like entering a new world.

Wallace was a fifth-generation descendant of Alexander Verdin and

Marie the *femme sauvage*. He had grown up on a section of the bayou that had once been the north end of the Indigenous community, but which after decades of erosion now represented the southern extent of the tribe's permanent presence on Bayou Pointe-au-Chien. Each of the bayou's original families had its own ancestral plot along the east side of the waterway—there was the Verdin family's land, the Naquin family's land, the Billiot family's land, and so on. Now, though, Wallace's family had grown too large for the tumbledown house on his family's plot. He and his wife had purchased another lot on the opposite side of the bayou and built a large wooden house, not grand but big enough for the Verdin family's growing brood. The local government had built a small concrete bridge across the waterway a few years earlier, which meant you could now drive from one side of the bayou to the other. Even so, the move was a major shift, and the Verdins were not the only family contemplating such a change: other bayou residents had started to move toward the mainland as erosion had advanced.

When Wallace was young, the bayou was a remote place, but now he felt more connected than ever with the mainland world. His elders had always traded with merchants from adjoining parishes, but now there was a paved road that ran straight from the bayou all the way up to the parish seat of Houma, and then on to New Orleans. It seemed as though the oil industry and the government that sponsored it had imposed their will on the bayou: the humble cypress houses came down and sturdier homes made of clapboard and brick popped up in their stead, along with double-wide trailers, and Wallace watched as cars displaced horses and pirogue canoes as the most convenient form of transportation.

The transition was not all a matter of choice. Erosion on the bayou had begun to destroy the subsistence farming and fishing that had allowed Wallace's parents and grandparents to live self-sufficient lives. The salt water that had once been absorbed by the dense thickets of the marsh now started to push farther inland with every high tide, worming through the perimeter islands that protected Pointe-au-Chien from storms and hurricanes.

The ground where Wallace's father had set their muskrat traps became so soggy that it was impossible to walk there even on dry days. The muskrats migrated farther inland and the family cattle started to get their hooves lodged in soaking mud. Even in places where the ground was still solid, the constant intrusion of salt water had made the old ways of life impossible— the backyard gardens that had once provided the tribespeople with fruit and vegetables now died out one row at a time, poisoned by the salt water. By the time Wallace's family moved to the other side of the bayou, the tribe could no longer rely on the land to keep them fed and healthy.

The changes on the bayou also forced more and more people to join the formal labor market, which for most people meant working for an oil company. In the decades since Texaco first arrived in Louisiana, the state's petrochemical business had grown so sprawling as to seem almost imperial, with hundreds of companies and thousands of jobs that ranged from tedious to backbreaking. The work could be grueling: one oral history mentions a man from the bayou who worked sixty-eight consecutive days on a rig in the gulf before he took one day off to go get a haircut. While at the barbershop, this man had the misfortune of running into his oil field boss, who ordered him to go relieve a colleague on a different rig nearby; the man then worked seventy-two consecutive days on the second rig before he got another day off.

Wallace's father had caught crabs and trapped muskrats for a living, and Wallace himself had spent much of his early adulthood raising oysters on leased sections of the marsh, but as his family grew he found he needed more reliable money than oysters could provide. Like almost every other man of his generation, he went to work in the oil fields, signing up for weeks-long rig shifts that took him west across the parish to Morgan City and sometimes farther along to Texas. The work was more lucrative than oysters, but it took Wallace away from his family, and there was a part of him that was disgusted to be working for the same business that had carved up the gulf. By the time the family moved across the bayou, Wallace had had enough—he decided to come back to the bayou and make a living on the water. The easiest way to do this, and indeed the only way, was to buy a boat and catch shrimp.

Inshore shrimping had sustained bayou families for generations, but over the course of Wallace's life it had blossomed from a subsistence practice into a bona fide career. Wallace's father had been the first man on the bayou to own a motorized boat, but now almost every man in the bayou had at least one Lafitte skiff, a ten-foot craft that was designed for shrimp trawling in the estuarine shallows. These men left every dawn by the dozen to go trawling, dropping their nets into the mosquito-pocked water as the sun came up and returning by midday to sell what they had caught.

In previous generations, most fishermen had caught only enough shrimp to feed their own families and perhaps their neighbors, but now they caught as much as they could and sold it to middlemen who came down from New Orleans. The total number of shrimpers in the parish never exceeded a few thousand, but the industry became essential for Pointe-au-Chien and other towns like it. Commercial fishing had a kind of symbiotic relationship with oil: when the price of shrimp fell one spring, more men would show up in the oil field the next winter, and when the oil market crashed, the laid-off drill men would turn to commercial fishing to make ends meet. Later on, several shrimp peeling plants opened on the edge of the water between these boat docks, large metal sheds where women would show up at all hours to peel and sort through thousands of pounds of fresh catch. The opening weeks of the May and August shrimp seasons became the most important times of the year, and bayou communities across the state began to commemorate the occasion with a unique ritual known as the Blessing of the Fleet—on the first Sunday of the season, the local priest stood on a bridge over the bayou and sprinkled holy water on the light blue Lafitte skiffs as they sailed out toward the marsh for their first catch.

Wallace's decision to start trawling for shrimp helped him escape the torpor of the oil field, but it also gave him an up-close look at the erosion process. He steered out along the bayou at the start of every new season to find that the marshland had changed in his absence: islands had migrated a few feet to the left, channels had grown wider or flattened out into brand-new ponds, this or that lake now took a few more minutes to cross. The bayou was always changing, thanks to the sediment inflow from the Mississippi, but it had never changed anywhere near this fast. The shrimp fishermen who trawled by moonlight often found themselves uncertain of where they were; they talked on the radio as they cast their nets into the water overnight, wondering out loud if the water would someday swallow the coastline whole. The land where they made their homes was still solid, still intact, but how long would that last? Like visitors from a distant land, they returned to shore bearing fearful messages, and those who had not seen the disappearance for themselves found the stories difficult to believe.

———

For Wallace's eldest son, Chuckie, the move across the bayou coincided with a move to a new school. After almost a decade of wrangling with the federal government, the state of Louisiana had at last agreed to integrate its public schools. Native Americans were classified as people of color under Louisiana law, so the tribespeople of Pointe-au-Chien had attended separate schools from their white neighbors. Wallace had gone to school in a chapel at the lower end of the bayou, where a Baptist group ran an Indian-only primary school to which most students traveled by pirogue canoe. Chuckie had attended first and second grades there, but now he would take a bus all the way up to the top of the bayou, where he would attend a school that up until that point had been only for whites.

Chuckie didn't understand why he had to switch schools, or why he would have to speak English at his new school. The north end of the bayou was only five miles away, but it felt like another world, and it even had a different name—Pointe-aux-Chenes, or "point of the oaks," a name that seemed to flaunt the mainland's immunity to the effects of erosion. As the bus rolled up toward the new school, the all-important bayou narrowed until it was just a ditch running alongside the road, too small and shallow for a boat to navigate. When they reached the parking lot, Chuckie saw dozens of adults gathered together, all of them white—they were parents who had come to protest integration. The children spent an hour on the bus before the sheriff arrived and forced the parents to stand down.

Today many residents on both sides of Pointe-aux-Chenes credit the integration of the elementary school with fostering new relations between the white and Indigenous populations of the bayou—as the two populations grew up together, they started to marry and have children, and the lines on the bayou started to blur. The integration of the school also drew Indigenous families away from Isle de Jean Charles, a nearby Indigenous community on a narrow island in the marsh. The remote island was losing more land to erosion every year, and once the school integrated, many island families moved to the mainland to escape frequent flooding. Even so, racism remained entrenched on the bayou for most of Chuckie's childhood: the teachers in the elementary school never seemed to treat Indigenous kids the same way they treated white kids, and when Chuckie started high school he often found himself getting in fistfights with white boys after class let out.

The one solace for Chuckie was his father Wallace's new job as a shrimp fisherman. He had never seen his father much when he was out in the oil field, but he idolized the work Wallace did trawling for shrimp, and he never had any doubt that he wanted to be a fisherman, too, when he grew up. When he didn't have to go to school his favorite thing to do was make the dawn shrimp run with his father—Chuckie would help steer the boat while Wallace worked the net and Chuckie's brother stood over the picking box and sorted through the catch.

When Chuckie was a teenager, Wallace bought a second Lafitte skiff, hoping to teach his sons how to trawl for themselves, but the boat's engine exploded one day while the boys were trying to start it, and Chuckie spent weeks in the hospital with burn wounds. In the aftermath of the explosion, Wallace decided he needed to move beyond small, unreliable skiffs—the boats couldn't run all the way out to the gulf without running out of fuel, and you could only use them during the summer shrimp seasons, whereas Wallace wanted to go shrimping year-round. He spent the next few years saving up to purchase a sixty-foot boat, an old steel trawler from Alabama.

By the time Chuckie graduated high school, Wallace owned multiple offshore boats and had hired a crew of part-time employees to run them along the Gulf of Mexico, so Chuckie took out a loan to buy the sixty-foot trawler from his father. He started trawling offshore as soon as he got out of high school, working with some of his cousins to haul shrimp out of the deep waters.

Chuckie had been shrimping for a few years when he met his first wife at a bar in Houma. She was Indigenous as well, but from another part of the parish. After they got married, the couple moved into a camper on the same property where Chuckie had grown up, next door to his father's house. Chuckie hoped to stay in the trailer for a few years while they saved up to build a full-size house, but his wife soon told him that she didn't feel safe living on the lower bayou. Flooding had only been a minor issue when Chuckie was a child, but now it seemed to be getting worse every year. The protective marshland barrier between the town and the gulf had all but evaporated, and now every time a hurricane or strong rainstorm barreled up toward the town, the surrounding grassland vanished beneath a torrent of storm surge. The bayou itself would flood backward and overflow, subsuming the old ancestral land beneath droves of water and flooding out the new backyard on the other side of the bayou where the Verdins had moved.

The flooding in Pointe-au-Chien was not like the flooding in the Keys or in Lincoln City, less an all-out blitzkrieg than a war of attrition: often the most significant impacts were psychological rather than financial. Many families had used recovery money after storms to elevate their homes ten or twelve feet above the ground, and now almost all the homes on the bayou seemed to be perched on stilts, except for Wallace's brick house and Chuckie's trailer.

Chuckie didn't want to leave the bayou at first, but at last he consented to look at a new house up in the town of Montegut, on a just-developed street called Aragon Road. It pained him to acknowledge, but he spent so much time out on the offshore boat that it didn't make all that much difference where he lived. He was at sea for three out of every four weeks anyway, and during his off weeks he would want to spend as much time as possible at home with his children. The house on Aragon Road was just a few miles away from the upper end of the bayou, about a twenty-minute drive from Wallace's house. The new house backed onto the much wider Bayou Terrebonne, a major waterway with drawbridges along it to let big ships through, so Chuckie was able to moor his boat in his backyard, something he had never been able to do in the narrow waterway of Bayou Pointe-au-Chien.

The couple moved in 1984, and the very next year a new storm arrived to confirm that their decision had been the right one. The erratic Hurricane Juan spun through the bayou, killing nine people on offshore oil rigs and destroying more than fifty thousand homes. One of those homes was Chuckie's camper, which the storm sent spiraling out into the marsh.

III.

Those Who Leave and Those Who Stay

Ten years after Chuckie and his family moved to Montegut, an old foe returned to the bayou. Louisiana Land and Exploration filed yet another trespassing lawsuit against the Verdins and their neighbors.

For as long as Louisiana Land had been on the bayou, the company had had to contend with the wiles of one Sidney Verdin, a great-great-grandson of Alexander and a distant cousin of Chuckie's. Sidney was born in 1910, and by the time he reached adulthood he had become an informal leader for the tribe, somewhere between an elected official and what the French-speaking bayou residents called a *nonc*—the literal meaning is "uncle," but on the bayou it connoted something like a community elder. Sidney maintained a small outpost a few miles south of where Chuckie had grown up, a little shack known in local lingo as a "camp." The area around the camp had once been the site of a permanent tribal village, but the incursion of the oil industry and the onset of coastal erosion had driven almost everyone away, and now only Sidney still ventured down into occupied territory to check on his shrimp nets and his small herd of cattle.

Sidney's camp sat in the midst of the extraction infrastructure erected by Louisiana Land's oil partners, sandwiched between an oil pipeline and a gas pipeline. Louisiana Land needed to control this territory at all costs, so it hired a private boat patrol to discourage intruders. Every day the patrolmen scoured a new section of the marsh in speedboats, stopping trappers to ask for their lease paperwork. For decades, Sidney Verdin and his comrades made frequent appearances in the company's patrol reports. In May of 1956, for instance, the patrolmen noted that they "found fence being built on company property"; the following year, after discovering the culprit, the patrolmen went back to the ridge "to serve injunction papers on [*sic*] Sidney Vardin [*sic*] who is trespassing on LL&E property," and spent some time waiting "to catch Sidney Vardin [*sic*] trespassing . . . but he did not show up." The following decade the company dispatched a helicopter to keep tabs on several ongoing construction projects in the area: in 1968, the aerial patrol logged that it flew over the camp "to check on possible trespassing by Sidney Verdun [*sic*]." The company had destroyed Sidney's illegal bridge only the previous week, but he had already rebuilt it.

Sidney's campaign against Louisiana Land had persisted into the 1990s, and now his son Gary had joined the fight. The family had continued to bedevil company patrolmen all through the eighties and nineties: one report from 1986 recounts that "our purpose this day was to pull out the walk bridges which had been built across our property line . . . we removed 5 bridges this day and stacked the lumber on Sidney Verdin's side of the ditch." The patrolmen had been accompanied by two deputies from the parish sheriff's office "in case of a possible confrontation." This was a reasonable concern, as there had been one occasion when Sidney and Gary had rammed their boat into a backhoe while the operators were trying to dig a canal. Indeed, Gary had grown up to be even more of a terror than his father: a favorite activity for him and his brother Alton was to speed on the local highway so that the police would pull them over, then beat up the officers and drive away. According to local lore, he once stood up in a courtroom after being sentenced for a misdemeanor, shouted that no one had the balls to take him to jail, and walked out scot-free.

Now the company was taking Gary to court for erecting another permanent fishing camp on company property, a green shack on stilts with a large shrimp net hanging off the side. The lawyers also accused Gary of tampering with its pipeline infrastructure. Gary and his codefendants countered that in fact it was Louisiana Land that was trespassing: the bayou belonged to the tribe by ancestral right, they said, and their acts of sabotage had been an attempt to stake the tribe's claim to it. LL&E dismissed this with acrimony. "The reality is that the defendants are more trespassers than possessors and their acts, if anything, amount to disturbances," its lawyers wrote.

Gary's lawyers tried to move the case to federal court on the grounds that the federal government had jurisdiction over issues relating to Indian land. The only problem with this argument was that the tribe's claim to the land had always been informal. Federal law grants ironclad legal protections to Indian reservations, but the Pointe-au-Chien tribe had never been recognized by the federal government, nor had the marshland ever been designated as tribal land. The residents of the bayou knew that they were the distant descendants

of the Chitimacha and Choctaw, and they had the years of racism and preju-
dice to prove it, but they had never organized themselves as a formal tribe. The
legal arguments could only go so far, and after a few months the judge handed
down a verdict against Gary Verdin, barring him from LL&E property and
ordering him to pay restitution for his sabotage against the company.

Chuckie Verdin was a distant relative of Gary's—the two men shared
a common ancestor—and it pained him to see Louisiana Land push his
cousin off the tribe's ancestral land. When Louisiana Land and Exploration
sliced up the marshland and bulldozed the tribe's old fishing camps, they
had severed the threads that linked the tribe to the bayou, bringing about
the end of a peaceful and independent way of life. Now they were bringing
the full force of the law to bear against the few remaining tribespeople who
wanted to defend that way of life. Dozens if not hundreds of tribe members had already moved inland over the years, Chuckie included, and the
community on the lower end of the bayou had eroded in tandem with the
marshland that surrounded it. The people who bought the vacated homes
were not locals from the bayou, but so-called weekend warriors, white
suburbanites who drove down on summer Fridays from the New Orleans
suburbs, lured by ample fishing catch and the presence of a large marina.
Chuckie worried that if the remaining tribe members did not defend their
claim to the bayou, the tribe's history would soon disappear altogether.

There were numerous reasons for the bayou's disappearance, and many
of these reasons were unique to Pointe-au-Chien—there was the upstream
damming of the Mississippi, the carving up of the marsh, the subterranean drilling in the water off the bayous, the endless succession of hurricanes, and the slow expansion of the oceans. Even so, the tribespeople of
Pointe-au-Chien were experiencing something that many other communities would experience as the climate crisis evolved: they were watching
their homes slip away as the environment around them changed. It didn't
happen in the span of one day like a wildfire or a hurricane, but over the
course of generations, slow enough that no individual member of the Ver-

din family had been able to see the collapse for what it was. Nor were the tribespeople of Pointe-au-Chien the only tribe confronting such a fate. The Gullah Geechee people of South Carolina and Georgia were watching their swampland homes erode away, the Quinault Nation in Washington State had been battered by storm surge, and several Inuit villages in Alaska had been forced to move by rising seas and thawing permafrost. These tribes had some of the oldest traditions in the United States, and they lived on some of the most vulnerable landscapes, but more protected communities would inevitably confront similar dilemmas.

Chuckie was never the loudest person in tribe discussions. In most settings he preferred to listen rather than talk, squinting out of a face weathered by years on the ocean. He also hadn't grown up learning the tribe's pidgin French or practicing its handed-down rituals, and he didn't spend all that much time thinking about the history of the bayou—in fact, he didn't even live on the bayou anymore. This is all to say that he was an unlikely candidate to lead the tribe's response to the Louisiana Land lawsuit. But it was he who would go on to become the tribe's longest-serving chief, and who would help the tribe organize its campaign for federal recognition.

As Chuckie saw it, the theft of the tribe's land had caused the outmigration from Pointe-au-Chien. If the tribe could get its land back, he told himself, then maybe the community would return as well. Maybe not everyone would move back down to the lower bayou, he thought, but at least more people would come back to visit family and reconnect with their tribal roots. They would be able to catch shrimp, row canoes, and visit the burial mounds just as their parents and grandparents had done, without fear of harassment by LL&E. The land would never again be fit for trapping or raising cattle, of course, but if the tribe could fulfill Sidney Verdin's quest to win it back, that would be enough.

Chuckie knew LL&E would never surrender the marsh to the tribe, not after all the trouble Sidney and Gary had caused over the years. He didn't trust the courts to do what was right, either. The only way to get the land back, it seemed to him, was to go through an entity even more powerful than LL&E: the federal government. The Bureau of Indian Affairs allowed tribal organizations to submit petitions for federal acknowledgment, seeking to be recognized as legitimate federal tribes. In the late nineties, as the lawsuit progressed, Chuckie and a few other tribe elders incorporated the Pointe-au-Chien Indian Tribe as a nonprofit and set to work proving what they had always known: that the land was theirs. Sidney Verdin was the tribe's first chief, but he passed away a few years after the tribal organization incorporated, and Chuckie took his place.

The federal recognition process is notoriously arduous, complex almost to the point of absurdity. An applicant tribe must prove not only that it "comprises a distinct community and has existed as a community from historical times until the present," but also that it "has maintained political influence or authority over its members as an autonomous entity from historical times until the present," and must also provide an exhaustive list of names and family lineages for living members. The United Houma Nation, a larger tribal organization in the same parish, had attempted the process a decade earlier, collecting reams of documentation and a member list of more than seventeen thousand names, only to see their petition rejected by the government after more than a decade of back-and-forth.

Chuckie enlisted his younger sister, Christine, a teacher who had once taught at Pointe-aux-Chenes Elementary School, and together the two compiled an exhaustive list of living and dead tribe members, more than six hundred in all. With the help of his family, the tribe's lawyers, and several sympathetic academics, the tribal organization assembled almost a hundred thousand pages of documentation showing the connections between the present-day bayou residents and their ancestors. The group filed for a separate petition in 1997 and sent its documentation to the government in stages over the next decade. In the meantime, Chuckie won the election to the position of tribal chief, and Christine was elected to the tribal council, while Wallace joined the council of elders.

The Office of Federal Acknowledgment is not a speedy organization—when the United Houma Nation had sought recognition, it had taken the government more than a decade to issue a verdict. In the case of the Pointe-au-Chien Indian Tribe, though, the government only engaged in "active consideration" of the petition for three years. In May of 2008, the office reached out to Chuckie to deliver its preliminary decision on the question of the tribe's identity: "The evidence in the administrative record," the government wrote, "is insufficient to demonstrate that the petitioner meets all seven mandatory criteria required for Federal acknowledgment." The tribe was not going to get its land back, at least not with the government's help.

—

For Chuckie Verdin, the pain of federal rejection was accompanied by an even worse indignity. Just a few years after the government sent back its finding, he had to give up shrimping.

Starting in the early nineties, the shellfish industry had exploded in developing countries like India, Indonesia, and China, all of which were catching millions of pounds of shrimp each year and selling them for cut-rate prices. The market price of shrimp crashed almost overnight around the turn of the century, falling from eight dollars a pound to two dollars a pound or less. The summer shrimp season had once netted Chuckie enough cash to get his family through the winter, but now he was barely breaking even. The collapse in shrimp prices altered the calculations for his trawling voyages, making it unprofitable to take the boats out unless he was sure of a big catch. Even if he caught droves of shrimp, he sometimes didn't make enough to pay his employees and earn back the cost of fuel.

Over the years, Chuckie had built out his shrimp operation into a successful small business: he owned several boats, worked with more than a dozen men, and caught thousands of pounds of shrimp a year. The price collapse forced him to lay off several of his workers and sell all but two of his six offshore boats, and he stopped using even those two boats as the years went on. The money had become so inconsistent that it was no longer possible for him to turn a profit, even if the catch in the gulf was good. The sixty-foot trawler his father had bought decades earlier stayed tied up behind his house on Bayou Terrebonne, aged and useless. The work that had sustained his father, grandfather, and great-grandfathers could no longer sustain him.

The bayou had changed, or the world had changed, and now Chuckie had to change, too. He got a job as the captain of a tugboat that ran up the Mississippi River. The job wasn't in the oil field itself, but he was still working for the oil industry: his boats carried benzene shipments from oil refineries along the gulf. These tugboat companies employed many former fishermen from the bayou parishes, and the work had a lot in common with the shrimp runs Chuckie had made out in the gulf—it took men away from their families for twenty or thirty days, then brought them back for a mere seven or ten. The only difference was that the money in the oil industry was better. You weren't working for yourself, but you could always trust the pay would be good.

Chuckie felt conflicted about working for the oil industry, but he had never been to college, and working on a tugboat was the best job he could get with the skills he had. The shifts took him up the river to Baton Rouge or all the way along the gulf to Bay St. Louis, Pensacola, Miami, up the Atlantic. It was easy work, boring work, and that made it all the harder to spend so much time away from home. There was nothing to keep his thoughts from straying back toward Pointe-au-Chien, nothing to keep him from thinking about the rapacious progress of the water. He was going on fifty years old, and he could not help but wonder what the bayou would look like in another fifty years. It was far harder for him to stop shrimping than it had been for him to move away from the bayou two decades earlier: he had spent the majority of his life's waking hours on a shrimp boat, and giving up the job felt more like a betrayal of his heritage than giving up his home.

Meanwhile, Chuckie's extended family, which had once been concentrated on the lower bayou, had begun to splinter off. He had been one of the first tribe members to move up to Montegut, but dozens of other families from the bayou and from Isle de Jean Charles had followed him there, including his parents. Even as many old bayou families moved northward, some peeled off and moved elsewhere in the state or out of state altogether. Chuckie's younger cousin Robert Verdin moved to central Mississippi after marrying a woman who had grown up there. Robert's brother Jerome followed him to the same town a few years later, hoping to escape the constant flooding down in Pointe-au-Chien. In the years that followed, two of Chuckie's other cousins, Leon Verdin and Elliot Verdin, moved to Alabama and Florida respectively. A succession of monster hurricanes over the next decade—Katrina and Rita in 2005, Ike and Gustav in 2008—pushed more people to leave the bayou.

A couple of years later, the BP Deepwater Horizon oil spill sent thousands of gallons of slick oil meandering along the Gulf, dealing a knockout punch to the beleaguered fishing community on the bayou. As the oil slick drifted closer to shore, the state government suspended the shrimp season, forcing even small-time fishermen to stay ashore and look for other work. Not only did the spill drive the lingering fishermen out of work, forcing them to give up the work that had sustained five or six generations of their forefathers, it also hastened the pace of erosion by killing off plants that held the marsh in place. Some of the most stubborn bayou residents went out shrimping anyway, but the market for their catch was far from what it had once been.

Despite his disgust with the forces behind the spill, Chuckie took some money from BP to spearhead the cleanup around Pointe-au-Chien, helping train the unemployed fleet of fishermen how to scoop up the rainbow oil in special nets. He would later serve as the plaintiff in a lawsuit against BP demanding compensation for the damage the spill caused to the bayou wetlands. A few years after the spill, he sold off his last boat, retaining only a small Lafitte skiff like his father's, something he could use to fish on his days off from the tugboat.

Chuckie's children, Angele and Charlie, grew up without the same connection to the bayou that Chuckie had. The family still sailed out to Chuckie's favorite barrier island in the summer, fishing and tanning on the edge of the Gulf, but the kids attended elementary school in Montegut, made friends in Montegut, and found summer jobs in Montegut. Chuckie spent twenty days out of every thirty on the boat in the gulf, and he never got around to telling his kids the stories he had received from his grandmother. The kids didn't grow up around basket weavers, didn't hear drums in the evening or watch people paddle pirogues back and forth past their house. They knew they were Indigenous, but the old tribal practices had faded over the previous generations, displaced by the arrival of the oil industry and the creation of new rituals like the Blessing of the Fleet. Wallace had not imparted the tribe's oral traditions to Chuckie and his siblings, and Chuckie in turn had not imparted them to Angele and Charlie. The family's traditions were their own, not those of the tribe.

Chuckie could tell early on that his son, Charlie, had a better knack for school than he had, and from a young age he hoped that Charlie would be the first in the family to attend college. Charlie earned a marketing degree from a nearby state school and started looking for jobs in sales and technology, but he soon found that there were no such jobs on the bayou or anywhere close to it. Almost everyone he grew up with went into the oil industry, and Charlie soon drifted apart from his school friends as they spent months away on tugboats and rigs. He started spending more and more of his time on the internet, striking up friendships with people who posted on a forum for his favorite video game. He and his friends started musing about starting a company, one that could handle merchandise creation for video game developers. The leader of the group lived in Tucson, and he invited Charlie and the other friends to join him out there and build the company together.

Charlie had never lived away from the bayou, indeed had never even considered doing so, but there was nothing keeping him there. He packed his bags and drove out to Tucson—dry, new-built, unfamiliar in every way. To his surprise, he liked it there (although he missed bayou food), and after a few years he invited some of his high school friends to join him and help with his burgeoning merchandise company. One friend came, and then another, and then another, until there were half a dozen Louisiana expats in Tucson working under Charlie's supervision. Like him, these expats had never really thought about moving out of the bayou, but there were no opportunities there. If you didn't want to work in shrimp or oil, you were out of luck.

Even as their extended family slipped away from Pointe-au-Chien, Chuckie and Christine continued to bolster the tribe's presence on the bayou. They cobbled together grant money for a tribal building right next to the house where they had grown up. The building was a steel Quonset hut elevated fifteen feet off the ground, with posters and billboards hung throughout that recounted the tribe's history and the progress of erosion on the bayou. Christine also started a summer "culture camp" so that the children of tribe members could reconnect with their heritage—the kids learned phrases in the tribe's pidgin French, practiced basket weaving, and took drum and singing lessons from tribal elders. In her spare time she continued to work with the tribe's lawyers on a revision to the tribe's acknowledgment petition, hoping to unearth new evidence of her ancestors' historical presence on the bayou.

These activities kept her busy, and at many times they helped keep her sane, but even she couldn't convince herself that her work for the tribe would draw people back to the bayou. The landscape itself had vanished, and so had the way of life that had sustained the tribe; that was not something you could just undo, not with any amount of money. The clearest proof of this could be seen right next to the new tribal building. The house she had grown up in, the one Wallace had bought when they were children, sat right in the shadow of the Quonset hut, but it had been battered into submission by years and years of wind and flooding. The house sagged on its foundations, and its door yawned wide and dark.

IV.

The Last Days of Pointe-aux-Chenes Elementary

It was March of 2021, and Sheri Neil was throwing together po'boys for the lunch crowd at her namesake Sheri's Snack Shack, the only restaurant in Pointe-aux-Chenes. The counter-service sandwich joint on the north end of the bayou stands about twelve feet off the ground, with a big red deck, where people can sit as they enjoy one of Sheri's renowned milkshakes.

At the height of the lunch hour, a woman drove into the parking lot and came running up the stairs. She was a teacher at Pointe-aux-Chenes Elementary School, which served about eighty children from both the bayou and nearby Isle de Jean Charles. Earlier that morning a representative from the parish school board had shown up unannounced and informed the staff that the parish was closing the school, effective that summer.

There were about a dozen people at the restaurant when the teacher drove up, and each of them ran at once to tell their families and friends. By nightfall everyone in town had heard the news, and by the next morning the residents of Pointe-aux-Chenes leapt into action as only the residents of a small town could. They started a Facebook group on behalf of the school and alerted the new cub reporter for the daily newspaper in the nearby city of Houma. Chuckie Verdin called the attorney who had represented the tribe in its federal recognition campaign and asked her to help them file a lawsuit against the parish. The town staged a small picket outside the school, with students and parents holding up handwritten signs.

This was far from the first school closure in the parish, which had seen broad population loss over the previous two decades. The story was more or less the same in every town: the shrimp business crashed, the flooding got worse, and people moved up to dry land, leaving empty desks in every classroom.

No one who lived in Pointe-aux-Chenes could deny that the bayou population was shrinking. The parish had shut down the library branch a few years earlier, warehousing the books in the school building, and the lower bayou had lost two grocery stores in the past decade. The only remaining general store, run by a woman from Isle de Jean Charles, was operating on thinner and thinner margins. You couldn't go more than a mile without seeing a FOR SALE sign.

Still, closing the school at this time felt like an unnecessary escalation, one that would push the town further toward depopulation and decay. Fifty years earlier, when Indigenous kids like Chuckie Verdin had first attended classes there, the school had been a hostile place for the tribespeople of the lower bayou, but in the decades since it had become a kind of cultural melting pot for the whole bayou community, a bridge between the white Cajun and Indigenous sides of Pointe-aux-Chenes. The school had one of the largest Native American populations of any school in the state, and teachers made a point of educating students about the rich history of the bayou, bringing in tribal leaders to demonstrate ceremonial dances and drum rituals. The bayou had no museum, no archive, no dedicated historian, so it was through the school that each generation of residents passed down their unique traditions to the next. If that went away, what would the town have left?

Even more painful was the fact that the decision had come just a few years after the Army Corps of Engineers had finished a new levee system that would protect the bayou, part of a massive project the agency had been working on since the aftermath of Hurricane Katrina. The erosion exodus that had begun with Sidney Verdin's generation and continued through Chuckie's seemed like it was finally about to slow down: the main reason so many people had left over the years was to escape the flood problem, but now the town would be protected from all but the most devastating storms. The marshland outside the levees might disappear, but the town itself would be safe for decades to come.

Just as the town of Princeville provided a foil to the buyout experiment in Lincoln City, another community near Pointe-au-Chien demonstrated what happened without investment in flood protection. A few miles to the west was the Indigenous village of Isle de Jean Charles, which like Pointe-au-Chien had been losing population for decades amid storm and erosion—around 98 percent of the island's landmass had disappeared. The federal government had excluded Isle de Jean Charles from its protective levee network, and rather than protect the island with flood walls the state government had opted to relocate its remaining forty-odd residents to a new tract of land farther inland. The relocation was funded by the federal government through an Obama-era grant program, and it amounted to the first whole-community climate migration in the history of the United States. The original idea for the relocation had come from a tribal elder, but many had grown dissatisfied with the state's handling of the program: the new site lacked direct access to the water that had sustained the island tribe for generations, and there were rumors that the local government might allow the uninhabited island to become a hub for vacation rentals and fishing tours.

The Isle de Jean Charles relocation represented yet another cautionary tale about climate migration, a lesson in how much culture and history stands to be lost when movement becomes a necessity. The residents of Pointe-au-Chien had hoped they would avoid this fate after the completion of the levee system—the most optimistic residents were saying the bayou was poised for a minor renaissance now that the state had addressed the main driver of migration. The closure of the elementary school dashed these hopes: Pointe-au-Chien might be better protected than Isle de Jean Charles from flooding, but in the long run it was destined to suffer the same cycle of disinvestment and depopulation. Levee or no levee, the outcome would be the same. Decades of erosion had changed life on the bayou for good, sealing the fates of both tribal communities.

The Terrebonne Parish School Board convened the next month to take a final vote on the closure. The meeting began with a public comment period during which parents and community members could address the board. The nine members sat Supreme Court–style at a long wooden desk, all arranged to face a single public podium. The residents of the bayou stood up one by one, white and Indigenous, and pleaded with the board to reconsider its decision. A few board members seemed moved by the show of support, but it wasn't enough: the board voted six to three to shut the school down. The eighty-odd students at Pointe-aux-Chenes Elementary would attend Montegut Elementary the following autumn. The lawsuit against the parish was still pending, but it didn't seem likely to succeed, since the board had the authority to manage its school system the way it saw fit.

Among the audience members at the meeting was Mary Verdin, whose husband was Alton Verdin, a tugboat captain and lifelong resident of the upper bayou. Alton's uncle had been the legendary Gary Verdin, and in keeping with the labyrinthine family trees of the bayou, Mary was Alton's fifth cousin on both his mother's *and* his father's side.

Working on a tugboat didn't bother Alton the way it bothered his cousin Chuckie. The pay had been enough to support Mary and their seven children all through elementary and high school, not to mention Mary's mother, who lived with them and helped them take care of the kids. The family had a one-story brick house on the upper end of the bayou, the part that was known as Pointe-aux-Chenes rather than Pointe-au-Chien and which had once been off-limits to Indigenous people like them. The wide marshland on the edge of their property sometimes flooded during heavy rains, but the house itself was modern and sturdy, and the family had hunkered down there during several hurricanes. Some of his older relatives still lived farther down the bayou, in the open-water areas that Sidney and Gary had called home, but much of Alton and Mary's extended family had moved up to join them on the solid territory of the mainland.

The school closure hit Mary hard, driving her first to depression and then to anger. Five of her seven children had graduated from the school already, but Gabrielle, the second youngest, still had one more year to go before she graduated to middle school, and Raelynn, the youngest, was just two years old. Mary had always been involved at the school, collecting box tops and Community Coffee proofs of purchase, and they lived close enough that she and Alton could go and have lunch with their daughters when Alton was home from the tugboat. One year Alton had driven his daughter Abigail to a father-daughter dance in a stretch limousine—the drive took, in total, about thirty seconds—and had shown off his traditional Cajun dance moves in the school cafeteria during the talent show. Now all of that would vanish. Gabrielle would finish elementary school in the ancient Montegut Elementary building one town over, with its steep stairs and single set of bathrooms, and Raelynn would never set foot in the school that had witnessed so much history.

To Alton, who had lived in Pointe-aux-Chenes his whole life, it seemed like the levee had arrived too late. With the school closed, the out-migration from the town would become all but irreversible. Who would move down the bayou to start a family, to raise their children, knowing that with every passing year a new rip would appear in the town's social fabric?

The closure of the school had started to make Alton and Mary doubt their future in Pointe-aux-Chenes. They needed to rip the floors out to fix long-term water damage, which would take thousands of dollars, and Alton wondered whether they should sell the house and find something inland in Montegut or Houma. Their eldest daughter had just become a real estate agent and was looking for her first commission, so she was helping them scout out houses that might serve as suitable replacements. Both wanted to move, but they didn't want to leave Pointe-aux-Chenes. Even as the school year began, they were stuck in a holding pattern, waiting for a sign about what they should do.

Gabrielle attended Montegut Elementary for less than two weeks before Hurricane Ida cut her school year short. The storm intensified to the threshold of Category 5 over the course of just three days as it pushed up the Gulf of Mexico, and made landfall a few miles south of Pointe-aux-Chenes with winds of around 150 miles per hour. The parish issued a mandatory evacuation order ahead of the storm, but many hardened bayou residents stayed behind and watched as the wind ripped telephone poles out of the ground and sheared the walls off double-wide trailers. The erosion of the bayou had eliminated the natural protection system that weakened storms as they made landfall, allowing Ida to retain its full strength for far longer than it would have decades earlier.

The devastation in Pointe-au-Chien was total. It took close to a week for the water to drain back out of the lower bayou, and when aid workers at last made it all the way down the length of the bayou road, they found that almost no structure had escaped the storm. It would take weeks for the parish to restore electricity and running water and months for FEMA to haul in the first temporary trailers, and even longer to drag away the mountains of gnarled debris that lined the side of every road. The sole remaining grocery store sustained so much damage that its owner, Mary's uncle, decided to shut it down for good. The final insult was that the storm had seemed to confirm the parish board's decision to shut down Pointe-aux-Chenes Elementary. The school in Montegut had survived the storm, but the old white building on the bayou had not. The storm had twisted the structure's metal roof like a nautilus shell and rolled it out into the street. There were shards of white wood all down the block.

Alton and Mary's house was in better condition than many of the trailers and elevated houses around them, but it was far from livable. The roof was in tatters and water had dripped into the bedrooms and the living room. Alton's contractor told him it would take about seven months to fix the house. In the meantime, he and his family would have to find somewhere else to stay, as would thousands of other people from Pointe-aux-Chenes and elsewhere in Terrebonne Parish.

It might sound counterintuitive, but the storm strengthened Alton and Mary's resolve to stay in Pointe-aux-Chenes. They figured if their house had survived Ida, it could survive just about anything, and they didn't want to abandon their ailing hometown as it began the tortuous recovery process. Unfortunately, it wasn't up to them: there was almost no livable housing anywhere on the bayou, and certainly none that they could rent on a short-term basis. The storm had walloped the nearby city of Houma, destroying dozens of hotels and apartment complexes, which meant the closest rental they could find was all the way in Mississippi. The owner asked for $900 a month at first, but by the time Mary went to go look at the place he had jacked it up to $1,500, plus a steep deposit. She said she'd rather buy a generator and take her chances back in Pointe-aux-Chenes. The following summer, as the residents of Pointe-aux-Chenes struggled to make it back to the bayou, the Louisiana state legislature voted unanimously to reopen Pointe-aux-Chenes Elementary as a French-language magnet school. The tragedy of the hurricane had inspired lawmakers to override the parish board's decision and offer the bayou community a new lease on life.

Alton, Mary, and the kids returned to their battered house once the power and water came back on, and Gabrielle resumed school at Montegut Elementary, taking some of her classes in trailers. Alton hoped that his insurance company would pay to bring a trailer to their property while they waited for the house to be fixed, but trailers were in high demand, and the insurance company was trying to haggle him down on the payout. All down the bayou, hundreds of other families were in a similar limbo. They couldn't come home, but they had nowhere else to go. The long process of displacement that had begun with the arrival of the oil industry and continued through an endless succession of floods was still going on, and Alton and Mary had no reason to think they had seen the end of it. Even once the school reopened, it would take a long time before Pointe-aux-Chenes got back to the way it was, if it ever did.

Nevertheless, the Verdins hunkered down. They would try to hold on a little longer.

CHAPTER FIVE

Frankenstein City

Houston, Texas

Two Hundred Acres on White Oak Bayou

I n late 1973, the United States plunged into an energy crisis. A group of oil-producing countries in the Middle East placed an embargo on exports to the US, causing petroleum prices stateside to skyrocket. The US was extremely dependent on foreign oil, and inflation made it difficult for American companies to ramp up domestic production fast enough to fill the supply gap created by the embargo. The sudden energy shortage rattled the nation's economy: hundreds of cars lined up at gas stations for rationed fuel, restaurants shut down when they couldn't afford to turn on their lights, and towns found themselves forced to ban neon signs and Christmas light displays.

There was one place in the United States where the oil embargo was good news, though, and that was Texas. The state had established itself over the previous decades as a major center of American oil production, and as the nation scrambled to achieve energy independence after the embargo, investors flocked southward to endow new wells and refineries. The natural landing place for this money was Houston, a midsize city best known at the time for its large shipyard and its proximity to a NASA base. In the space of just a few years, the city ballooned from a regional outpost into a boomtown. By some estimates, more than a thousand people arrived there every day, chasing a new glut of jobs in drilling, land surveying, metal fabrication, equipment sales, corporate strategy, and more. The ramp-up in oil production minted hundreds of overnight millionaires, swaggering businessmen who showed up drunk to board meetings and raced around town in brand-new convertibles. Traffic on the freeway got so bad that oil executives started flying helicopters to work, and soon the local airport needed a dedicated air traffic controller just to handle the morning rush hour.

As Houston swelled to accommodate hundreds of thousands of new arrivals, developers gobbled up land on the outskirts of town. The city had been inching out mile by mile from the downtown core, annexing a new slice of land every few years, but now the sprawl stumbled out in every direction. The price of land spiked, old families on the outskirts of the city sold their ranches to developers, and builders sketched out subdivisions spiraling out along the spindle spokes of future highways. There were no zoning restrictions, no neighborhood master plans, no attempts to design a city that looked any specific way. The city just built. Hundreds of new neighborhoods wrestled for space, each with its own anodyne name—Aberdeen Trails, Hearthstone Park, Copperfield Southcreek Village, and so on. Some developers laid their streets out in a simple grid, others chose curlicues and winding avenues, but for the most part it was hard to tell the subdivisions apart. They jostled together cheek by jowl, claiming every inch of undeveloped prairie.

The history of most of these subdivisions has been unremarkable, but there was one that was destined for something beyond mere middle-class stasis. In the years just before the oil boom, a real estate developer named Larry Johnson had purchased a few hundred acres of farmland at the northwest edge of the city. It was plain pasture, lined with unpaved roads and bisected from west to east by the muddy waters of a stream known as White Oak Bayou. As the local economy took off in the seventies, Johnson's company sectioned the land into a six-part subdivision he called Woodland Trails West. He and his business partner sketched out winding streets and cul-de-sacs and lined them with one-story ranch houses, giving the development a distinct suburban feel. The company even incorporated a homeowners association for the new residents, complete with its own set of rules—no lawn signs, no loud noises, and no landscaping unless approved by the "Architectural Control Committee."

The first residents of Woodland Trails West were mostly Houston new-comers, middle-class families that had come into wealth on the heels of the boom and purchased a spot in one of the city's newest developments. The neighborhood was well-to-do, quiet, and white. The houses were single-story brick affairs, modest enough, but they had wide lawns and large ga-rages, which was just about all the new residents could ask for. The property taxes were reasonable, crime was nonexistent, and the downtown center was only a half-hour drive away on US 290, the brand-new freeway. The neigh-borhood's secluded position, set back from a main commercial road, meant that there were seldom any outsiders passing through.

There was just one wrinkle in this picture of suburban bliss. Running alongside the edge of the neighborhood was White Oak Bayou, which drained from the northwest prairie out to the Gulf of Mexico. For most of its history the bayou had been a muddy and oft-flooded creek, but by 1970 the local government had tamed the stream in preparation for future development. Acting in accordance with the engineering wisdom of their day, county officials razed the trees that surrounded the waterway, dredged a channel for the water, and lined the sides of the channel with mounds of concrete. The goal was to limit the scope of potential flooding, but the ef-fect was to encourage developers to build right up to the brink of the creek.

Larry Johnson and his fellow developers might have guessed that their new development would flood someday, but it didn't matter. There were no zoning laws in Houston, nor were there any real restrictions on permitting. The county government had never conducted flood studies of the bayous and officials had no interest in limiting development in those areas, not with the city going through such boom times.

By the early nineties, though, the residents of the neighborhood had begun to feel that the bayou was a little too close for comfort. When Johnson and his fellow developers built atop the area grassland, they covered up fertile soil that excelled at soaking up excess rainwater. Since it could no longer seep into the ground, this rainwater now sloshed down streets and sidewalks. The developers who carpeted Houston with concrete had been running up a flood debt with Mother Nature. Every time they built a new neighborhood on the northwest edge of the city, they destroyed more of the natural drainage system that had helped prevent flooding. The rainwater that fell on the uplands still had to drain southeast through the city and into the gulf, but now there was no soil to soak it up as it passed through the landscape.

This debt came due around the turn of the century in a massive one-two punch. The first blow was Tropical Storm Frances, which filled up the retention pond near the neighborhood and pushed several feet of water onto the streets, flooding more than four hundred homes in the area. A mere three years later came Tropical Storm Allison, a fragmented and devastating storm event that skulked over Houston for almost a week and left almost no part of the city untouched by flooding. The storm's vortex retained its strength long after making landfall thanks to a phenomenon known as the "brown ocean effect," wherein evaporating soil moisture provides extra fuel to a storm that might otherwise fizzle out. This meteorological quirk made Allison a kind of natural prototype for later climate-enhanced storms like Hurricane Irma, which drew extra strength from the warm waters of the Caribbean, and Hurricane Ida, which maintained its strength as it passed over the waterlogged marshland of Louisiana. Even though the storm never achieved hurricane-force winds, it still dumped around twenty inches of rain on the city, enough to submerge whole neighborhoods and waterlog thousands of cars on the city's web of freeways and tollways.

The whole city had suffered under Allison, but the flooding in Woodland Trails West was different. The water had risen as high as six or seven feet along many of the houses, filling the neighborhood up like a bowl. The neighborhood's location and the geography of the surrounding area made Woodland Trails West a natural destination for excess water from the bayou, which meant the area would flood not once in a generation but every time the bayou crested its banks.

The crisis confronting the residents of Woodland Trails West was far from specific to Houston. It was happening in hundreds of other cities and suburbs around the country, anyplace where builders had torn up swamp grass and thrown down concrete. The rapid outward pace of real estate development had slammed up against the immovable facts of nature, creating urban and suburban landscapes that faced perennial risk from flooding. The residents of these neighborhoods were the victims not of bad luck but of bad design.

—

A few years before Larry Johnson's firm sketched out the plans for Woodland Trails West, in 1968, President Lyndon Johnson signed a bill creating the National Flood Insurance Program, which offered homeowners government-sponsored flood coverage at a discount. The program also represented the federal government's first attempt to sponsor a "managed retreat" from flood-prone areas: since it was mandatory for people who lived in flood zones to buy insurance, the program amounted to a de facto tax on living near the water, and thus softly encouraged people to move elsewhere.

To implement the law, though, the government first had to define who lived in a flood zone. This was easier said than done, since floods are messy and unpredictable events. No matter how wide you drew a flood map, a future flood could always go past it. The officials who designed the program did not resolve this problem so much as lean into it: they chose a nice round number, the "hundred-year flood," as the national standard for how far out a flood zone boundary should extend. The term "hundred-year flood" was something of a misnomer: it was not a flood that happens once every one hundred years, but a flood that has a 1 percent chance of happening in each year.

It took about a decade for the federal government to finish drawing these flood maps for the entire country. When the maps for Houston were complete, surrounding Harris County adopted them as building guidelines: starting in the mid-1990s the county barred new construction anywhere in the hundred-year floodplain, hoping to shift developers away from waterways and toward higher ground. By then, of course, it was too late: there were already thousands of families living in areas that the new flood maps showed were too risky, including Woodland Trails West.

The only way to protect these families from the risk they faced was to undo the sins of the oil boom by making space for the water. In the aftermath of Tropical Storm Allison, the county ramped up a home buyout program that operated on the same principles as the one in Lincoln City. The county used FEMA money to buy property owners out of their homes, move them elsewhere, and destroy the old homes, with the aim of turning the neighborhoods into natural water sinks. The difference in Houston was sheer scale: Woodland Trails West alone contained around as many residents as Lincoln City had, and there were dozens of similar neighborhoods that also got buyouts.

Just like in Lincoln City, buyouts in Houston were a matter of accounting. The homes in Woodland Trails West were destined to flood again, and the National Flood Insurance Program would have to keep paying to fix them every time, so why not dole out one big payment and be done with it? In other communities, ones that were more historic or less vulnerable, this might have been a difficult decision, but this was Houston—there were hundreds of identical communities in the city, many of them just down the road. But just like in Lincoln City, buyouts were also a *social* policy, a way of deciding who should live where and how far the government should go in helping its citizens obtain adequate shelter. They took place in an unequal city, in an unequal economy, and they did not resolve that inequality so much as perpetuate it.

The demographic makeup of Woodland Trails West when the buyouts began in 2000 was very different than it had been when the neighborhood first opened. Over the previous thirty years, northwest Houston had experienced a prolonged episode of white flight: when the price of oil cratered in the mid-eighties, it plunged the city into a recession, which in turn triggered a wave of foreclosures in middle-class havens like Woodland Trails West. By the time the recession ebbed, another wave of new suburban housing had already been built farther out from the downtown core, drawing wealthy white families away. In the meantime, the city had become a destination for thousands of immigrants from Mexico and Vietnam, many of whom moved into the now-affordable subdivisions. Many of the original white residents still lived in the neighborhood, but their neighbors no longer looked like them, or at least they didn't *all* look like them. Woodland Trails West had become that rarest of things in a contemporary American city—a diverse community.

The flood buyout after Tropical Storm Allison cleaved this community apart along lines of race and class. As the white and Hispanic residents moved away from the neighborhood, their pathways through the city diverged.

Cindy Mazzola was one of the first to leave. A quiet, thoughtful woman with a fondness for jigsaw puzzles, she had lived in the neighborhood for around fifteen years by the time it started flooding. She and her husband had moved there right after they got married, but the two divorced in the years between Tropical Storms Frances and Allison, and her husband moved out, leaving Cindy to see their two teenage sons through high school. Cindy was an analyst at the computer company Compaq, and she made enough money to keep up with the mortgage on her own, so she didn't have any immediate plans to leave. All the boys' friends lived in the neighborhood, so she didn't want to let the divorce get in the way of their stability.

The experience of floating out of the neighborhood on an inflatable life raft forced her to reconsider those plans. She put the family's one-story ranch house on the market just a few months after Allison, but she couldn't find any buyers, and when the county approached her with a buyout offer, she figured she had better take the bird she had in hand. Her sons protested at first, disconsolate at having to leave the neighborhood behind, but they reconciled themselves once they saw all their friends' families taking buyout offers, too.

Cindy searched the city for another neighborhood with the tree-covered charm of Woodland Trails West, hoping to ease the transition, but she ended up finding something even better. It was a community called Winchester Country, with Western-themed street names like Pony Express Road and a large clubhouse in the center. The family's new house was bigger than the one they had left behind, and it was only a year old when they moved in, plus there was room in the backyard for Cindy to add a small swimming pool. Perhaps the best part was the steep driveway, which sloughed rain down to the street before it could gather in the yard.

This was how the buyout process was supposed to work, and indeed it was how it worked for most people in situations like Cindy's. She had paid off most of her mortgage in the old neighborhood, so she could afford to buy a substantial house somewhere else, and the salary she drew at Compaq allowed her to keep up with the pricier loan.

A few months after moving into Winchester Country, Cindy was surprised to see one of her old neighbors from Woodland Trails West getting into his car just down the street. There were dozens of subdivisions within a five-mile radius of the old neighborhood, and yet they had both happened to choose this one. They weren't the only ones, either: FEMA records show that at least three other people moved into Winchester Country from Woodland Trails West, and that at least ten other people from the old neighborhood moved into the surrounding postal code.

This was not coincidence so much as demographic destiny. When Cindy had moved to Woodland Trails West, the neighborhood had been the very image of the suburbs: majority white, family-oriented, and situated on the very outskirts of the city. By the time she left the neighborhood twenty years later, it was majority Hispanic, with housing that was old by Houston standards, and the city had sprawled out so far that it no longer qualified as a suburb. Whether they intended to or not, Cindy and her former neighbors had moved into a neighborhood that looked the way Woodland Trails West *had looked* when they first moved there. The buyout funds had allowed them to turn back the clock: by moving a few miles farther outside the city, they moved backward along the demographic timeline.

This trend fits within the larger pattern discovered by the researchers Jim Elliott and Kevin Loughran, a pair of professors at Rice and Temple Universities who have conducted some of the only longitudinal studies of flood buyouts in the United States. Elliott and Loughran used federal and local records to track down thousands of Houstonians who took buyouts after Tropical Storm Allison, hoping to identify patterns in where they ended up.

The researchers found that buyout participants from wealthier areas were more likely than those from low-income areas to move to a nearby neighborhood, and furthermore that participants from wealthier neighborhoods were more likely to resettle in proximity to *each other*. That was what had happened to Cindy. On the other hand, the researchers found, buyouts that took place in low-income areas often scattered residents much farther afield, as well as far apart from each other. This made sense on an intuitive level: people who had the means to choose where they moved would tend to move close to where they had lived before, choosing similar neighborhoods and upgrading when possible. Those who did not have those means would have to look far and wide for an affordable house, or would maybe even leave the city to live with friends or family.

Woodland Trails West was about half white and half Hispanic, and it was neither very wealthy nor very poor, so a bit of everything happened after the buyout. There were many people who moved back to their original hometowns outside Texas, or moved to stay with family in Wisconsin or Michigan, never to come back; one couple moved all the way to Northern California, to a trailer park that later burned down in a wildfire. Almost all the white residents I spoke with followed a similar path as Cindy, moving northwest into an outer belt of suburbs that included Cypress and Spring. The demographic shifts in the neighborhood played a role in these decisions: one resident told me that she left because she hated when her new Mexican neighbors had played music at odd hours, and Cindy told me that as the school district had diversified, her son had witnessed more fights and gang activity in his high school.

Among Hispanic residents in Woodland Trails West, the outcome of the buyout seemed to depend on when it took place. The immigrant families who had lived in the neighborhood longer tended to end up in the same new-built suburbs as their white neighbors—the equity they held in their homes and the extra boost provided by the buyout gave them enough leverage to upgrade just as Cindy had done. The story was different for more recent arrivals to the neighborhood. These families had less equity in their homes, they received less money from the government, and they had less leverage to decide where they would move.

This was the case for Audelina and Brigido Gomez, who moved into Woodland Trails West after Tropical Storm Frances in 1998, but before Tropical Storm Allison in 2001. The family had lived in an apartment complex close to the neighborhood for several years while they saved up to buy a house where they could raise their two young daughters.

Woodland Trails West seemed to Audelina like the ideal place to buy a first home. It was set back from the main road, the streets were quiet and shaded from the large trees that hung overhead, and there were plenty of parents and young children playing outside on afternoons and weekends. They found a house just a few dozen paces from White Oak Bayou, a brick three-bedroom with powder-blue garage doors.

Audelina's real estate agent asked the seller about flood risk, but the seller didn't mention any issues. The entire neighborhood had flooded only the previous year during Frances, so this was at best a lie by omission, but under state law the seller of a home had no obligation to disclose flood history or flood risk to prospective buyers. The seller had to disclose if there was lead paint, asbestos, or a fire history, but they could omit flooding. Because Audelina and her husband didn't speak much English, they weren't able to get any information from the white residents who had lived through the previous year's flood. The first time she learned about flooding in the neighborhood was when Allison rushed through in 2001 and sent residents wading out to the highway through several feet of rising water.

Audelina knew she had been deceived by the seller of the home, and her realtor tried to help her sue the seller's realtor for damages, but the lawsuit went nowhere. When the verdict came back against the family, their realtor suggested that Audelina look into the flood buyout program—almost everyone else in the neighborhood was planning to enroll.

Audelina and Brigido weren't sure at first whether they should stay in Woodland Trails West. They had dreamed of living somewhere like this, and their daughters had just entered elementary school in the school district. Plus, they had just bought the house, and it was on the cheaper side for the area, so they weren't sure where else they could afford to go. Even as they pondered their decision, though, the subdivision started to disappear around them. People like Cindy were already gone, and now even those who weren't sure they could afford to upgrade were taking the buyouts, either because they were anxious about flooding or because they didn't want to be in a half-empty neighborhood. Those who weren't participating in the buyout put their homes on the market and moved elsewhere, bringing new families like the Gomezes in to assume the risk that came with living in the neighborhood. A cloak of silence settled over the streets, so that sounds started to echo all the way down the block. The only people who came to the neighborhood were the Red Cross charity workers who stopped by every now and then to deliver meals to construction workers repairing homes, and even they tried not to stay too long. Woodland Trails had become an unnerving place, like a nuclear test site or the ruins of some failed utopian community.

Brigido and Audelina had only made a few payments on their mortgage, and they still owed almost the entire cost of their home. The buyout from the county would zero them out on their existing loan, but they would have almost no money left over for a down payment on a new house. The county provided no additional funds to facilitate the cost of the move, and indeed the buyout agreement required them to vacate the house within weeks of signing the paperwork. The Gomezes didn't have the time for a prolonged house hunt, and in the last few days before they moved they found themselves scrambling to pack up the few belongings that had survived the flood.

The only solution was to downgrade, to purchase a much cheaper home somewhere close by. The kids could stay in the same school, the couple's commutes would be the same as they had been before, and they wouldn't go bankrupt on the new mortgage. The family found a small subdivision of close-packed lots just a mile away, squeezed between a couple of warehouses, and they bought a two-story house on a street called Poncha Springs Court. Their neighbors in the new area were almost all recent immigrants: the subdivision had been built only a few years earlier, and by that time the surrounding area had transitioned from mostly white to mostly Hispanic. It was nowhere near as idyllic as Woodland Trails West, and there was nowhere at all for Audelina's daughters to play—you couldn't walk for more than five minutes without entering an industrial parking lot—but at least they were safe from the water.

Or so they thought. Audelina had made sure from her new neighbors there was no flood history in the subdivision on Poncha Springs Court, but she had not known to check the area's position on the new floodplain maps the county was starting to use. It turned out that the neighborhood was just as vulnerable to White Oak Bayou as Woodland Trails West had been, and that it, too, had been built in the stream's natural floodway. The houses in the neighborhood sat a little higher above the street than did the ones in Woodland Trails West, and the driveways sloped down to drain rainwater, but the Gomezes were still sitting ducks for future flooding. The county had not placed any restrictions on where buyout participants could use their money, nor had officials monitored where the Gomezes chose to move, so there was nothing to prevent them from moving out of one flood zone and into another.

It took a few years, but eventually a hurricane passed over Poncha Springs Court, and the Gomezes' new street started to flood. Audelina and her family watched from the window as it rose past their mailbox, and then up their sloping driveway, and then started to seep into the garage. The family barricaded themselves in the bathroom—Audelina and her two daughters sat shaking in the bathtub, and Brigido paced back and forth across the tile. The water peaked in the garage and never entered the house, but the family no longer thought of themselves as safe. They had chosen to sacrifice the comfort of Woodland Trails West in exchange for the safety of the new neighborhood, but once again they had been duped. Even as many of their former neighbors beat a retreat to the newer, safer suburbs, they and hundreds of other families remained behind in the aging neighborhoods closer to the city center, higher off the ground but still in the water's path.

———

The goal of the county's buyout program was to turn Woodland Trails West into a reservoir for White Oak Bayou, letting the area revert back to nature, but to do that the county had to buy out every last home in the subdivision. That was easier said than done.

The first problem was that some residents liked the seclusion of the half-cleared neighborhood. The county knocked down all the homes it bought, but it continued to mow the grass on the lots where those homes had been, which meant that those who stayed behind were now surrounded by acres of open space. For residents who had several cars and trucks, or whose children liked playing football, or who appreciated the seclusion from the noise of city life, this change incentivized them to stay. There were even reports elsewhere in the city of families who took a buyout on their home, bought the house next door, and then flipped the second home for more money by advertising its large "side yard."

For low-income residents, the problem with the buyout was that they had nowhere to go. Houston was experiencing another housing boom, and many families in Woodland Trails West knew they would struggle to afford a mortgage anywhere else in the city. The options were either to stay and try not to think about the flood risk or to drive farther away from downtown until they reached a neighborhood that was more affordable. If they had a job nearby or if their kids were enrolled in the local schools, the decision to stay was easy.

A third group of residents bedeviled the county officials by selling their homes on the private market rather than to the government. The county's buyout effort was the largest ever attempted in the United States, and the FEMA program itself was still new when Allison struck, which meant that buyouts often took three to five years to execute. Many families couldn't wait that long, or they figured they could make more money by selling to underinformed buyers like the Gomezes. They pawned their homes off to new residents, moved somewhere safer, and left their successors to deal with the consequences. This was unethical, but it was also the inevitable outcome of the state's flood disclosure policy—since there was no requirement to reveal flood history, it made no financial sense for a seller to deflate the market value of his or her home by revealing that it had flooded.

The buyouts in Woodland Trails West tapered off after the first few years, but the county kept chipping away at the neighborhood over the course of the next two decades. The remaining homes passed from hand to hand as buyers flipped them for more money or rented them out to short-term tenants who came and left. By 2021, almost all the residents who lived in the neighborhood during the storm had left. In their place lived an assortment of holdouts and new arrivals who couldn't or wouldn't leave the neighborhood behind. The new Woodland Trails looked like a more manicured version of Lincoln City, with the same eternal silence but none of the debris and overgrown trees. Someone had installed a disc golf course on one of the old blocks, attracting enthusiasts from around the city who seldom stopped to ask why the neighborhood was so spacious and empty.

Other neighborhoods along White Oak Bayou and the parallel waterways of Houston's urban core had also been bought out and destroyed, leaving isolated pockets of empty land scattered throughout the city's inner ring of developments. It had taken twenty years of bureaucratic effort, but the county had begun to transform Houston's built environment, rubbing out the ill-conceived developments of the first boom and carving out space for the water. The city had become a Frankenstein's monster, a patchwork creature ripping off pieces of itself and grafting them onto other parts in an effort to adapt to the constant risk of flooding.

Thousands of buyout families, meanwhile, had slipped into Houston's larger currents of development and demography, following larger patterns of white flight and sprawl. The same rain had fallen on all of them, but they lived in an unequal city, and had met unequal fates.

II.

Squeezing the Lemon

J ust as every boom is followed by a bust, so every bust is followed by a boom. The oil crash of the eighties had shocked Houston's economy, but the city roared back to life as it entered the new millennium, buoyed by the fracking revolution and droves of new arrivals from other states and countries. The city's population surged until it rivaled that of Chicago.

Even as the county government worked to undo the floodplain sins of the first boom, developers resumed their outward stampede. The city billowed out like a mushroom cloud, swallowing thousands of acres of prairie, papering over rivers and streams, unfurling ribbons of asphalt. Its tentacles reached out to undisturbed towns and villages, swept away centuries-old farms, crossed county lines, and kept going.

When the federal government published its first flood insurance maps, Harris County restricted development in the hundred-year floodplain, forcing developers to build new houses farther away from the water. This was an admirable policy, but it did not prevent developers from building in flood-prone areas, in large part because the hundred-year flood threshold was an unreliable metric. The government calculated the size of this area by looking at historical rainfall data for a given area and identifying a benchmark storm event, but these historical records were often flawed or incomplete, and, in many places, the government did not even have a hundred years' worth of good data.

The result was that there were many, many more vulnerable homes than there were within the boundaries of the hundred-year floodplain. The county's efforts to control flooding were only as good as its estimates of flood risk, and those estimates were flawed—they accounted neither for the rapid pace of development in Houston nor for the new reality of climate-enhanced storms. As county officials tried to undo the mistakes of the past, they remained blind to the mistakes of the present. It was only so long before another storm came along and opened their eyes. That storm was Hurricane Harvey, a monster cyclone that stalled over Houston in August of 2017 and dropped more than fifty inches of rainfall on the city over the course of three days, changing the calculus for what we can expect from twenty-first-century storms. Harvey was a thousand-year flood event, many times more severe than anything the county had ever incorporated into its plans. The storm exhibited all the strange and terrifying characteristics common to hurricanes in the age of climate change: it intensified to full strength in a matter of days, stalled over the moist gulf landscape, and seemed to contain an infinite amount of rain. Even for seasoned Houstonians, it was truly unprecedented.

If the story of Houston's risky development had been written in invisible ink, Harvey was the lemon squeezed on top of it, revealing a truth that had been there the whole time. The pattern of flooding across the city illuminated the neighborhoods and subdivisions where developers and local officials had underestimated or ignored climate risk. It was in those areas that thousands of people lost their homes to the water, and it was in those places that the churn of displacement soon began.

Some of the flooding was predictable. Woodland Trails West flooded again, as did other buyout tracts throughout the city's inner core. In these places, Harvey only further underscored what the county already knew, which was that many older sections of Houston should never have been built at all. The years after Harvey would see the county redouble its buyout efforts in these areas, leveraging new federal money to acquire thousands more homes in the neighborhoods it had already been trying to buy out. The county also targeted new areas within the hundred-year floodplain, many of which had escaped its attention after Allison: the Meyerland neighborhood, a middle-class white community on the city's southwest side that had seen repeated flooding in the years before Harvey; the Allen Field area, a low-income Hispanic community adjacent to Greens Bayou where most of the houses were already elevated on wooden stilts; and a series of large apartment buildings along the city's south side whose parking lots and courtyards had filled with filthy water.

There was another layer to the Harvey flooding, though, and this one was far more troubling. The flooding occurred not only in the older developments of the inner ring but also in newer communities farther away from downtown, most of which had been built well after the debut of the FEMA flood zone maps. These neighborhoods sat outside the hundred-year floodplain, in places where there were no building restrictions, so residents of these areas had not had to purchase flood insurance. Harvey seemed to be one of those unaccountable disasters—courts later found that the storm qualified as an "Act of God," an event that "could not have been reasonably expected or provided against." It had flooded even those places where water was not supposed to go.

Upon closer inspection, it became clear that God had nothing to do with it. The real problem was that the flood zone maps were far too narrow, which meant that many subdivisions in supposedly safe areas had been destined to flood from the start.

Take for instance the neighborhood of Wimbledon Champions Estates, a luxe development on the far northwest side of the county, near a large river called Cypress Creek. The homes in the neighborhood were two-story colonials, bland but stately, most boasting three-car garages and ample backyard space. The street names were arch references to the greats of tennis—McEnroe Match Drive, Agassi Ace Court, Borg Breakpoint Drive. The developers had built the first stages in the late nineties and kept building through the next decade, persisting through the setback of Tropical Storm Allison and the county's new policy of retreat from the bayous.

Wimbledon Champions was not in the hundred-year floodplain, which meant that there were no legal restrictions on what developers could build, and the residents of the neighborhood were not required to purchase flood insurance. The neighborhood was, however, in the *five-hundred-year* floodplain, a broader category encompassing the storms that had at most a 0.2 percent chance of happening each year. Almost the entire neighborhood flooded after Harvey, displacing hundreds.

A few miles to the east there was a neighborhood called Timarron Lakes, nestled inside a massive community called The Woodlands. The risks in this neighborhood were less the result of hubris than outright manipulation of the federal map system. The developers had wanted to build in the oxbow of a waterway called Spring Creek, but the land they sought to use was in the hundred-year floodplain, so they had devised a workaround. They purchased several hundred tons of dirt, hauled the dirt over to the side of the creek, and dumped it on the ground, raising the ground level in the neighborhood just a few inches above the hundred-year floodplain threshold. The developers then wrote a letter to FEMA asking the agency to grant a "letter of map revision"—in other words, to gerrymander the maps—so that they could build on the site. The dirt wasn't enough to stop Harvey, though, and after the storm, residents returned to find their Mc-Mansions filled with brown water.

Then there were the neighborhoods that sat outside any kind of floodplain and away from any major bayou, those places where the issue was not riverine flooding but rainfall drainage. Houston's drainage grid was ancient and inadequate. The sidewalk drains and storm grates were far too narrow to accommodate a massive rainfall event like Harvey, and indeed they had failed several times in the past during lesser storms; many portholes had filled up with leaves and debris over the years, and the city had fallen behind on servicing them, especially in low-income neighborhoods close to the city center. The apocalyptic precipitation of Harvey had overwhelmed the system within the first few hours, but the rain had kept falling, and soon the water had risen so high that it started to enter residents' homes. The problem was worse in places like the historic Black neighborhood of Sunnyside, where more than two-thirds of residents saw some form of flood damage. The county had spent billions of dollars on protecting residents from river flooding, but it had failed to prepare for the devastating effects of severe rainfall on impervious concrete. The flooding was not only in the hundred-year and five-hundred-year zones, but also in neighborhoods that weren't in any kind of flood zone at all. The storm had revealed a truth that applied not just in Houston but around the country: urban areas were by their very nature vulnerable to climate-fueled rainfall.

Once again, Harvey forced planners in the Houston area to rethink their relationship with nature. The rapid onset of climate change had widened the gap between expected risk and actual risk, and thousands of people were caught in the middle. The entire fabric of the city seemed to be vulnerable to water, from individual homes and apartment buildings to highways and streetscapes and storm drains. Even with a war chest of money from FEMA, the county had only been able to reshape a small section of this landscape. Harvey had smashed that paradigm, flooding tens of thousands of homes that were not inside a floodplain or near a major waterway, and the county did not have a plan for the people who lived in those homes. They were on their own.

III.

The Churn

Even as Harvey neared landfall on the Texas coast, having steamed up to Category 4 strength on its way across the Gulf of Mexico, Becca and Sergio Fuentes still had no idea that the storm was about to devastate their lives. They were both native Houstonians, and they were used to the emotional roller coaster of hurricane season, but they also had bigger things to worry about on the weekend the storm approached. Becca was thirty-one weeks pregnant with twins, and she had a condition called vasa previa that made her pregnancy dangerous. She was supposed to go to Texas Children's Hospital the following week for bed rest and induced labor.

Becca and Sergio first met in middle school, and by August of 2017 they had been together for fifteen years. Becca was a teacher at a special-education school, and Sergio had done a series of mechanic and engineering jobs—he had worked on an oil rig, had repaired combat equipment in Afghanistan, and was now an electrical technician. Even with their combined incomes, it had taken them several years to save up for a down payment on their first home and a few more years to afford all the furniture.

Their house was in Bear Creek Village, a large and leafy neighborhood that sits about twenty miles west of downtown Houston. Bear Creek is farther outside the city than Woodland Trails West, and it feels even more secluded and suburban. The homes in the neighborhood are a perfect fusion between modernist grace and seventies-era quirkiness: most of them have jutting roofs or atypical, angular facades, and the best look like they could have been designed by modern architects like Eero Saarinen or Robert Venturi. The school district is excellent and there's an elementary school in the neighborhood—many kids walk to school without having to wait at a single stoplight. Becca and Sergio had chosen the house because it seemed like they could raise a family there over the long term. "Our plan was like, that was gonna be our house, forever," Becca told me.

But the neighborhood also happened to sit just west of a massive undeveloped forest known as the Addicks Reservoir, one of two reservoirs that protects downtown Houston from flooding. These reservoirs function as large natural bowls, collecting rainwater as it moves downriver toward the gulf. On an ordinary day, they're forest parks, but during large storm events they turn into temporary lakes.

Becca and Sergio lived less than a mile from the western edge of Addicks, but they had no reason to fear the reservoir, at least as far as they knew. They had never heard of any flooding in their neighborhood, and had never seen so much as a puddle on their street. The house in Bear Creek was the first they had bought together, and it was sometimes tough to meet the mortgage payments, so they hadn't purchased flood insurance—not only could they not afford it, but they also weren't in a flood zone, so no one had brought it up when they applied for a mortgage.

Despite the approaching hurricane, the couple had been looking forward to a relaxing weekend at home—they had waited out bad storms before, so they figured they'd be fine. The rain came down all day Friday, and all through Friday night, and it came down even harder on Saturday, but the storm drains seemed to be holding up, and there was still no water pooling on their street. On Saturday evening, Becca's feet began to swell and she felt herself overwhelmed by nausea as her labor approached. She alternated between sitting in the bathtub to settle her stomach and crouching in the hallway during the tornado warnings that arrived every few hours. Sergio checked on her every few minutes and also kept tabs on the couple's five dogs. The rain kept teeming down the whole time, hammering on the roof, spraying along the length of the street.

Late in the afternoon on Sunday, as Sergio was watching the news, a spokesperson for the Army Corps of Engineers appeared on the television. The Corps managed the Addicks Reservoir, which along with its counterpart to the south was filling up with water, and fast. Never in the history of the city had the reservoirs come close to reaching their capacity, but Harvey had slowed down over Houston, and the Corps was getting worried. There was not enough room in the reservoirs to catch all the rain the storm was expected to bring, and the agency couldn't release water out of the reservoirs without flooding homes on the other side. The spokesperson on TV said that everyone in Bear Creek Village needed to leave the neighborhood the next morning. If they didn't, they would be stuck in the neighborhood when Addicks overflowed.

Becca and Sergio loaded up their truck the next morning with as many suitcases as they could fit around the dogs. They and hundreds of other Bear Creek residents left the neighborhood and drove a few miles away to Becca's mother's house, where they waited for news. Becca's labor pains kept mounting over the course of the day, but she couldn't make her appointment at the hospital—the entirety of downtown Houston was flooded, along with several major highways. Monday was Sergio's birthday, and in the late afternoon he got a call from some friends who had navigated into Bear Creek Village on a speedboat. They had pushed open the door of Sergio and Becca's house and seen all their furniture floating inside.

It took until Friday morning for the water to drain out of downtown. Becca made it to the hospital later that day, four days late for her appointment, and the staff put her on bed rest. Becca's usual obstetrician had lost her home to the flooding and couldn't come to the hospital to check on her, so a temporary staff managed her labor. The new doctors told Becca that the twins were still very premature, and they wanted her to hold out for as long as she could before giving birth.

After Sergio dropped Becca off, he drove back west and met up with Becca's brother and a few of his friends. The police still weren't letting anyone back into Bear Creek Village, and they had barricaded the main entrance to the neighborhood, but Sergio couldn't keep sitting around—he needed to see the house. He and his friends gathered up all the flashlights and rain gear they could find and walked into the neighborhood through a side street. The water had not yet drained out, and in some places it was up to Sergio's chest. It smelled strange, and there was debris swirling all around it—Sergio kept bumping into submerged mailboxes as he walked. When he got inside the house, the stench was overwhelming. The furniture was soaked and flipped upside down, the air felt thick with mold, and a few feet of dark, stagnant water covered the entire house, laid over the ground like a mirror.

The next Wednesday morning, once the water receded, Sergio started working through the wreckage of the house, ripping out carpets and tearing down walls. He hauled hundreds of pounds of furniture out into the street, tore water-stained cabinets down from their moorings, and riffled through the nursery for every piece of moldy baby clothing. The deeper he moved into the house, the worse the destruction became. Mold had climbed up the walls and along the ceiling, infesting everything from chairs to stuffed animals to Tupperware.

Sergio checked in with Becca on FaceTime early in the afternoon. A few minutes after the couple hung up, the doctors told Becca that they needed to deliver the twins at once—she was still weeks away from her due date, but one of the boys' heads was pressing against an important vein, and the delivery couldn't wait. The doctors conducted an emergency C-section the next hour as Sergio rushed down to the hospital. He arrived just in time to meet his two sons, Marley and Hendrix, as they entered the world.

Now that the boys were safe and healthy, the couple could turn their attention to the other overwhelming fact in their lives: they had lost everything they owned. The flood had consumed their entire house, erasing all their assets and advantages. They hadn't lived in a flood zone, but still the water had wiped them clean. How was that possible?

The answer was that, just like Woodland Trails West, Bear Creek Village had been destined to flood. If you look at a satellite map of Houston, the Addicks and Barker Reservoirs appear as green trapezoids inside a gray city, the only untouched nature in a haze of development that has otherwise consumed every inch of land. The Corps of Engineers first laid plans for the reservoirs in the late 1930s after a pair of ferocious storms pushed ten feet of water through Houston's burgeoning downtown core. The young city was fifty miles inland from the gulf, so it was safe from the worst effects of ocean storm surge, but the Corps needed to control flooding downtown if the city was ever going to prosper.

When you consider Houston's subsequent history of unchecked development, the fact that the Corps built the two reservoirs at all is a remarkable concession to nature. The agency annexed valuable farmland to create a natural moat for the rest of the city. As with all the Corps's projects, though, the reservoir project hinged on a crucial assumption about flood risk: the agency acquired only enough land to handle rainfall from a thirty-inch rainstorm. The engineers knew that larger storms were possible, and their designs acknowledged the fact that such storms might fill up the reservoir and push water onto private land, but the engineers assumed such events were unlikely and did not think it worthwhile to buy up all the land in the "maximum flood pool." At the time, the reservoirs were ten miles west of Houston's farthest outskirts, and the private land in the flood pool was mostly pasture, so even if the reservoir did overflow, the runoff would only endanger cows and corn, not people. After the first oil boom, however, it became clear that someday the city would extend past the flood pools, and the Corps soon issued a warning to future developers: "As the surrounding areas are developed, this may mean that homes in adjacent subdivisions may be flooded," its engineers wrote in 1986.

That line turned out to be prophetic, and indeed it was something of an understatement. Climate change had created the conditions for even stronger hurricanes than the Corps engineers would have dared to imagine. It was not that "homes in" Bear Creek Village flooded. The entire subdivision flooded, every home and street and yard and park and driveway. For almost a week the neighborhood functioned as an extension of the reservoir, invisible beneath a glaze of taupe-colored water, only a few roofs and treetops poking out here and there. The neighborhood and all its inhabitants had fallen into the abyss between expected and actual risk, just like the residents of Woodland Trails West.

Except this time the cavalry was not coming. The federal government arrived soon after the disaster to dispense aid money and low-interest loans, but even FEMA did not have the resources to restore all the homes that Harvey had destroyed. If the neighborhood had been inside a flood zone, the National Flood Insurance Program would have paid out claims to homeowners, and the county might have arrived later to offer buyouts to the remaining residents, but the residents of Bear Creek received neither flood insurance payments nor buyout offers. Their best hope was a pending class-action lawsuit against the Army Corps, which alleged that the agency had built a reservoir that was destined to flood private property. Such a lawsuit would take years to resolve, though, and most people in Bear Creek did not have that long to wait.

Becca and Sergio moved into Becca's mother's place, assuming they would be back in a few months. The house was a godsend for the couple, but it was also a scene of total chaos: they were living there not only with the twins and Becca's mother but also with Becca's brother, his wife, and their two children, plus the dogs. Becca worried about how the frenetic environment would affect the twins, and she wanted to get back as soon as she could, so Sergio spent long hours trying to fix up the house in Bear Creek, tossing out thousands of dollars' worth of furniture and equipment, gutting out every molded piece of wood until only the studs were left.

By the end of the year, their plans had changed. Sergio had been offered the chance to open a Matco Tools franchise, a rare opportunity that would require a lot of upfront investment, but would pay off down the line. Sergio could take the franchise and Becca could quit teaching to help him manage the books, but only if they got out from under their mortgage in Bear Creek. They put the half-rehabbed house up for sale in early 2018 and received eight offers in a few days, many from investors and rental companies hoping to pick up properties on the cheap. One of the highest offers came from a family, so Beeca and Sergio went with that one—it felt poetic, since they themselves would never get to raise their children in the house.

The couple no longer had their mortgage payment to worry about, but they had emptied their savings on the flood recovery, so they couldn't even afford to get a lease on a new apartment, let alone a house that would accommodate the twins and the dogs. They stuck it out at Becca's mother's house for another six months, and then another year, and then another two years, praying that Sergio's franchise would start to gain some momentum. The coronavirus pandemic pulled the rug out from under them once more as Sergio's clients vanished and his usual sales route turned into a ghost town. Becca picked up a little money from an independent yoga practice, but it wasn't enough to make up the difference.

Four years after the storm, Becca and Sergio were no closer to finding a permanent home. They sometimes drove out to the suburbs on weekends to go to open houses in the new subdivisions or at cheaper condo complexes, taking the twins along with them to nearby parks. It was a nice escape from the chaos of Becca's mother's house, but they couldn't delude themselves that the tours were anything more than window-shopping. Even so, they knew they couldn't stay with Becca's mother forever. The twins would start kindergarten the following year, and Becca and Sergio still had no idea where they would enroll. They thought they might be able to start renting an apartment if business picked back up at the Matco franchise, but it stayed stubbornly below pre-pandemic levels. Another mortgage was out of the question: not only did they not have enough for the down payment, but the housing market had gotten so hot that they couldn't compete with most buyers anyway. The couple seemed to have fallen into a permanent limbo, a state of confusion symbolized by the storage locker they still rented a few miles away. They had received thousands of charitable gifts after the storm, furniture and clothes and supplies, but they didn't have space for any of it at Becca's mother's house. The baby clothes had sat in the locker so long that Marley and Hendrix had long since outgrown them.

Becca and Sergio never visited Bear Creek, and they seldom thought about the old house, with one exception. A few years after the storm, one of Sergio's customers stopped him as he was making a stop on his sales route.

"Are you the same Sergio Fuentes who used to live in Bear Creek?" the man said, and rattled off Sergio's old address. Sergio said yes, and the man said he lived in Sergio's house now. His parents had bought it after the storm and fixed it up for him and his wife. The man had recognized Sergio's name from some old junk mail that had shown up in the mailbox.

Sergio laughed at the coincidence, but it also gave him a disturbed feeling he couldn't shake. Someone just like him was living in the house now, sitting right in the path of the next flood, holding a mortgage that might at any moment turn into an albatross of devastating weight. The storm had taken away the life he and Becca had built together, leaving them with more responsibility, but with less than ever to show for it. Like countless other families in Houston, they had fallen into a churning vortex of displacement and instability. Government and market forces alike were buffeting thousands of people from one home to the next, moving them from one risk zone to another. The churn had swept them off their feet, and they still had no idea when they would hit the ground.

—

Becca and Sergio weren't the only people in Bear Creek who wanted to get out from under their mortgage. A horde of speculators and investors descended on the neighborhood in the weeks after the hurricane, snapping up every damaged house they could find. These investors ran the gamut from wealthy local individuals to large hedge funds, but they all shared a common tactic: buy flooded homes for cheap, fix the damage, and rent them out to newcomers. After collecting rent for a few years, the investors would make back the bargain-basement cost of purchasing the home, at which point they could either keep renting it out or sell it to someone else. This tactic had become common across the United States after the recession, and not only in places like Houston. Even outside the context of a hurricane, many Wall Street firms thought they could make some quick money by renting single-family homes to families who couldn't afford to buy them. The investment firm Blackstone had purchased thousands of these homes through a subsidiary and had at one time become the nation's largest owner of such properties. An analysis by the *Houston Chronicle* found that between five thousand and twelve thousand flooded homes were sold after Harvey, and that many of those sales were to large institutional investors and hedge funds.

Among the largest buyers that year was a company called Cerberus Capital Management. The firm was founded by a billionaire Donald Trump ally named Steve Feinberg, whose watchword in all things was secrecy—"If any [employee] at Cerberus has his picture in the paper," he once told shareholders, "we will do more than fire that person. We will kill him." Cerberus had around $50 billion in assets by the time Hurricane Harvey struck, with ownership stakes in a security contractor called DynCorp and the Remington Arms Company, and in recent years the company had also invested in single-family rentals around the country. The going had not always been smooth—its properties in Memphis racked up code violations at a much higher rate than those of other landlords, and the company was quick to file eviction notices when tenants were even a little behind on rent —but nevertheless the company grabbed up all the houses it could get after Harvey. By the end of the year, it had purchased more than a thousand properties throughout the Houston area. The firm would later bundle and sell many of those properties to other investors, but it retained a presence as a landlord in neighborhoods like Bear Creek, where it branded itself as FirstKey Homes.

The clients were people like Juan Torres, who started renting a house in the neighborhood in 2019; he had been living in the suburb of Humble, but he moved to Bear Creek to be closer to his job working with petroleum compressors. When the pandemic caused mass layoffs at his workplace, he transitioned to landscaping and property maintenance. Juan was satisfied with FirstKey Homes: he had never had any issues with flood damage, he told me, and the management company had been quick to fix his roof when a rogue tree branch fell on it. He thought the rent was fair given that the house was two stories and came with a two-car garage, but it was getting expensive—FirstKey had raised the rent from $1,520 to $1,650 that year, and he wasn't sure whether more hikes were coming. He didn't want to rent forever, and he had called FirstKey several times to see if they would consider selling him the house, but the company wasn't interested in letting go of the property, so Juan had kept paying.

Just as the county buyout had cleaved Woodland Trails West apart along racial lines, the investor frenzy in Bear Creek also had a racial dimension. The residents of the neighborhood before the storm had for the most part been older, white, established families who had moved to the neighborhood on the tail end of Houston's first boom; young couples like Sergio and Becca Fuentes were the exception, but they were getting more common as older residents moved away. Bear Creek was farther out on Houston's pinwheel of sprawl, and about a decade behind on the white-flight timeline, but it was following the same trajectory. The swarm of investor activity sped up the process, bringing about a decade of demographic change in the space of a few months. New arrivals tended to be younger and lower income than the former owners, and they were far more likely to be Black or Hispanic. The neighborhood now looked something like Woodland Trails West had looked in the late stages of the buyout: there were a few older white residents who had chosen to stay, and the neighborhood around them had exploded with diversity.

"The typical seller was someone that has lived in the neighborhood for a very, very long time, they had children there, and they were now empty nesters, and they didn't want to spend [years] trying to put the home back together," said Dustin Gaspari, an energy investor who purchased and rehabbed several homes in Bear Creek after the hurricane. "It probably in many respects accelerated some folks' intention to move out of the area."

That was true in more senses than one. Many longtime residents see this new diversity as more a curse than a blessing: they told me the neighborhood had gone "downhill" since the storm. Many speak in coded language about "renters," and more than a few of them told me that the changes *since* the storm had caused them to consider leaving, even though they had spent years working to rehab their homes after Harvey. The initial shock of investor activity had led to a series of white-flight aftershocks that continued even after the flurry of home transactions died down.

The social stratification between old owners and new renters sometimes erupted into outright suspicion or hostility. This was the case for Jacob Lee, a machinist in his thirties who moved into the neighborhood with his wife and young children around a year after the storm. Jacob and his wife rent a property owned by Cerberus Capital Management.

It was early in the winter of 2021, and another rainstorm had descended on Houston, causing the streets in Bear Creek to flood throughout the afternoon. Jacob was heading to pick up his kids from school, but as he drove down the neighborhood's main boulevard, he realized the water was getting too deep for his car. He pulled into a driveway, the first one he saw, and stopped there to wait for the water to drain out.

That's when he saw an older white man come out of the house across the street from the one where he'd parked, red-faced and shouting. The man was yelling at Jacob.

"Get out of his driveway," the man screamed. "Stay away from his house!" Jacob, who is Black, replied in as calm a voice as he could manage.

"I'm not backing down there," he said. "It's going to ruin my car."

"All right then," said the man, "I'm going to get my gun." He stormed off across the flooded street and emerged from his house a few moments later holding a pistol. Jacob locked the doors of his car and called the police.

The man stalked away from him and raged at passing cars, banging on their hoods, and yelling at them for splashing water onto his yard. By the time the police arrived, the man had gone back into his house, and a group of officers surrounded the front door, urging him to show himself. The man never emerged, but his wife came out and negotiated with the officers until they left. The police told Jacob that since they hadn't seen the man point the gun at him, they couldn't do anything.

Jacob passed the man several times on the street in the months that followed, but he didn't seem to recognize Jacob as the person he had threatened. In all likelihood, Jacob thought, the man had never understood that he had been aiming the gun at one of his neighbors. The churn that followed the storm had caused Bear Creek to transform around him, and he had failed to reconcile his image of the old neighborhood with the reality of the new one.

IV.

The Widening Gyre

If Houston was the site of a long duel between water and concrete, Hurricane Harvey represented the most decisive victory yet for the forces of nature. Even so, it was far from the end of the story.

After the storm, the county once again set about reshaping Houston's urban environment, correcting past mistakes just as it had done after Tropical Storm Allison. The county received billions in new money from FEMA and the federal Department of Housing and Urban Development in the first few years after the storm, and Houston voters also approved a billion-dollar bond for buyouts and other projects. Engineers widened channels, carved out new retention ponds, and started demolishing other pockets of the city. The county also revised the floodplain regulations to restrict new development in the five-hundred-year floodplain, expanding the zone of retreat. The Army Corps even promised to fix the Addicks and Barker Reservoirs to protect people like Jacob Lee from future flooding, although it wasn't clear how exactly the Corps was going to do that, since Bear Creek sat on land that would always flood if the reservoirs overflowed. Like thousands of other people in the Houston area, the residents of the neighborhood were living in a zone of permanent risk.

Meanwhile, another wave of short-distance migration began after Harvey, tracing the same contours as the wave that followed Allison. Just like their predecessors from Woodland Trails West, the residents who departed Bear Creek moved farther out on the city's matrix of sprawl, riding the same economic and demographic trends that had pushed millions of other Houstonians out into the fast-transforming prairie.

Medy Onia and her husband, Carlos, were a perfect representation of the multicultural metropolis that Houston had become. They had both been born in the Philippines, but they met in Nigeria on an international teaching fellowship. After a few years traveling the world, they decided to take a teaching fellowship in the United States: their children were approaching school age, and they wanted to ensure they got a quality education. They found their way to Houston, a city Medy came to love almost at once. It was affordable, it was diverse, it was warm, and it offered the right combination of excitement and privacy.

The Onias' home was just steps from the edge of the Addicks Reservoir in Bear Creek Village, so they knew they were vulnerable even before Harvey. Their home had flooded twice in the two years before the storm, first by a few inches in the Memorial Day floods of 2015 and then only eleven months later during the so-called Tax Day flood of 2016. Both of these events were classified as hundred-year storms—once-in-a-century disasters had become annual affairs. The Onias had finished fixing the damage from the latter flood just a few months before Harvey destroyed their home all over again. Medy is a bustling woman, and her tiny frame seems to contain infinite energy, but in the aftermath of Harvey she fell into a nervous depression. She stopped sleeping, cried at random intervals, and vowed she would never go back to Bear Creek. It was time to move on.

The Onias' path out of the neighborhood was very different from Becca and Sergio's, and a lot easier. The couple had signed up for flood insurance after they flooded in 2015, so they got a substantial payout after Harvey. They looked for homes in the city of Katy, a major suburb just a few miles to the west, and soon found a pristine new subdivision with winding roads of stucco houses, many still under construction. The new home the Onias bought there was bigger than the quirky seventies house they had occupied in Bear Creek, with plenty of room for their grandchildren and extended family. Thanks to the cash they earned from selling their Bear Creek home to an investor, they were able to afford it without much trouble.

Medy wasn't concerned about the fact that their new house abutted a channelized waterway that snaked through the city of Katy. She also didn't worry about the fact that the land their home occupied had once been an ancient prairie, an enormous natural sponge, but most of the grass had been shredded by developers as they rolled out subdivision after subdivision along the length of Interstate 10. The waterway had at one time been a rambling stream, but the county had long since engineered it into submission, turning the creek into a massive ditch around fifteen feet deep. Even as Medy and Carlos moved into their new house, a construction team was busy fortifying the channel. Plus, their street was built in a style now common to Houston subdivisions—the homes sat on elevated mounds of dirt, and the driveways sloped down to the street, where there were storm drains of ample size.

When Medy went for a walk through the neighborhood she could see the waterway behind a row of identical houses across from hers. Even when it rained, the water never got anywhere near the top of the ditch, and she believed she would never live to see it rise that far. At the same time, though, she knew that there was nowhere in Houston that was safe from climate change. For as long as the city kept sprawling, and for as long as the warming atmosphere kept amassing and releasing new deluges of rain, flooding was not a matter of if but when. The worse the flooding got in the old subdivisions of the inner ring, the more people would move outward toward the exurbs like the one Medy and Carlos now called home. The white-hot housing market that followed the first years of the coronavirus pandemic only hastened this outward push, jump-starting another wave of growth in Sunbelt cities like Houston. The more people moved out to those suburbs, the more prairie grass would become concrete, and the worse the flooding would get downstream. The cycle of expansion and transformation recalled the language of W. B. Yeats's famous poem "The Second Coming": "Turning and turning in the widening gyre . . . Things fall apart, the center cannot hold." Medy and Carlos had not escaped the city's self-cannibalizing cycle, but perpetuated it.

Houston saw several catastrophic climate disasters in the years after Medy and Carlos left Bear Creek. Just two years after Hurricane Harvey, the city cowered beneath Tropical Storm Imelda, another torrential rain event that dropped more than thirty inches of water on the city, eclipsing Allison's record once again. Less than two years later, a ferocious ice storm descended on Texas, propelled south by a climate-enhanced fault in the global jet stream; the storm shut down the state's power grid, burst thousands of pipes, and killed more than forty people in the county. Later that summer, downtown Houston flooded again during tropical cyclones Nicholas and Pamela, not to mention during several no-name rain events that did not even make the national news. All these disasters spared the suburb of Katy, and Medy still believed the new neighborhood would be safe from flooding for the duration of her lifetime, but somewhere deep down she knew the reprieve wouldn't last forever. Someday the storm drains on her street would fill and brim over; someday the creek would crest its banks.

The battle was over, but not the war.

CHAPTER SIX

Why Should This a Desert Be?

DROUGHT, AGRICULTURE,
AND THE ERA OF EXPANSION

Pinal County, Arizona

When the Cows Didn't Come Home

T he summer of 2021 was long and dry, and Karen Felkins's cows were running away.

Karen and her husband, Bob, managed Diamond B Livestock, a cattle ranch in central Arizona. For more than fifteen years the couple had raised sport steers for rodeo shows: they bought a few dozen cows at a time from a cattle auction, raised them until they were old enough to perform, and leased them out for a few weeks at a time to rodeo owners across the state. The steers ran around the stadiums in thunderous stampedes, bucked off death-defying cowboys, wowed their audiences, and then came back home to the ranch for some well-deserved rest.

Their ranch was just outside the saloon town of Florence, about an hour southeast of Phoenix. The town sat close to the bends of the trickling Gila River, and from atop a long mountain ridge it overlooked a wide stretch of desert scrub, land untouched by anything except tire tracks and hoofprints. The Felkinses owned more than five hundred acres and leased tens of thousands more, which gave their cattle plenty of room to graze—a grown steer consumes as much as thirty pounds of grass per day, and the herd needed to be at full health to keep up with the demands of the rodeo stage. As the cows roamed out to the east, they could munch on the shrubs and tough weeds that grew up in the desert silt; when they got thirsty, they could wander down to the banks of the Gila and drink all the water they needed.

Things had begun to change that year. The heavy winter rains that were supposed to feed the desert grass had never arrived, and the early summer rains had not arrived, either, which left the landscape in an almost lunar state of desiccation. The state of Arizona, along with the rest of the West, had entered the twenty-first year of an interminable drought, an event that scientists believe has no real precedent in the last twelve hundred years. The endless succession of dry years had begun to scar the natural landscape in ways that brief rainstorms could not fix. The Gila River dried up, deprived of melting snow from the nearby mountains, and the scant shrubbery that dappled the desert shriveled up and died. Karen's cows trampled back and forth across their territory looking for something to eat and drink, but they found nothing but silt, whipped up into swirling dust clouds by intermittent gusts of wind. Karen had noticed them start to look slower and more emaciated, skinny enough that you could see their rib cages through their hides.

Before long, the cows went rogue. They rammed up against the wire fences that delimited Felkins's property, busting through the barriers one at a time in search of sustenance. Roving across the outskirts of Florence, they found their way into a field of irrigated alfalfa that belonged to one of Felkins's neighbors, at which point they began to chow down. Karen and her husband rounded up the steers, drove them back to the ranch, and patched up their rangeland fences, but the couple were shaken. They had never seen cows behave that way before.

With no natural pasture for the cows to eat, and no surface water for them to drink, the Felkinses had to pay out of pocket to keep the herd alive. The couple bought bales of hay from nearby farmers and arranged for a water company to bring them hundreds of gallons of water per day on a mobile tanker truck. It was around five times more expensive to feed and water the cows this way, and the costs quickly became unsustainable. Monthly bills for water and silage exceeded the amount of money the couple would receive for leasing the cows out to rodeo companies, and that was going to remain true even if the pandemic disappeared overnight. To make matters worse, the drought had forced farmers in the area to cut down on the amount of hay they were growing, which caused the price of the feed to soar even further. The Felkinses now had to compete for food not only with other ranchers but with the large dairies in the county, which depended on huge bundles of hay and silage to feed their penned-up herds.

By early summer, Karen and Bob had no choice but to downsize. They picked out the smallest, skinniest-looking cows, loaded them onto a trailer that they hitched to Bob's truck, and drove them two hours south to the Marana Stockyards outside Tucson, the site of a weekly cattle auction. The stockyards were packed and the air was grim: hundreds of other cattlemen and dairy owners had made the same calculations as Karen and Bob, and now it seemed like almost every rancher in the state was trying to off-load their cows, whether to save on the cost of feeding them or just to make back a little money where they could. Now the glut of cows had caused the price to crater, and many ranchers were walking away with pennies.

Ranching had always been an up-and-down business, dependent on inscrutable trends in customer demand and vulnerable to the caprices of commodity markets. Raising rodeo cattle was even more unpredictable. Karen had always viewed the industry as cyclical, defined by booms that enticed inexperienced newcomers and busts that cleared the field of all but the hardiest ranchers, but this bust seemed different. The state had gone more than eighteen months without significant rain, and there were no signs of relief.

Bob passed away on the Fourth of July that summer after contracting a short and sudden illness, leaving Karen alone to manage the herd. The drought was worse than ever, and without other options, she downsized further. Every weekend she loaded half a dozen cattle into her trailer, drove them up through the mountains, and sold them off in the town of Show Low, a ranching village on the border of the Apache reservation. There again she found many of her fellow ranchers selling off what they could, and some of them told Karen that they were getting out of the business altogether. Ranching was hard work, and it was only seldom rewarding. The younger generation wasn't interested in taking over. The drought was making life sick and miserable for cows and ranchers alike. What was the use of pushing through?

As Karen descended the mountains back to Florence, the ranchers' message rang in her head. She wasn't sure how much longer she could go on.

———

The first chapters of this book focused on the most severe and destructive climate disasters—sudden, extreme events that displace people from their homes and catalyze chaotic episodes of migration. The two most damaging disasters are floods and wildfires, which can destroy thousands of homes and uproot entire communities in the span of a few hours. Droughts are a different kind of disaster, more like the erosion that transformed Pointe-au-Chien—they are slow-moving, almost epochal, and indeed they only become disasters once they last long enough to seem almost normal.

By the same token, the disasters discussed in the earlier chapters of this book were local events, but droughts are regional. They change the conditions of possibility for entire states and ecosystems. The ongoing drought in the American West, which has warped the region for the past two decades, is the precondition that makes other disasters possible from California through Nebraska. If it were not so dry in California, the vegetation on the forest floors of the state would be far less flammable, which would lead to fewer wildfires. If there were more moisture in the ground, it would help absorb summer heat, which would make events like the Pacific Northwest "heat dome" of the summer of 2021 much less likely. The West has always been a dry place, but its ecology also relies on a bare-minimum water cycle, and the present drought has all but halted that cycle.

The gruesome effects of this prolonged dry spell had become apparent across the West by the time Karen Felkins's cows started to run away in 2021. In the Klamath River of Northern California, for example, the water fell so low that the summer sun could heat the river to scalding temperatures, boiling salmon alive in their streams. Millions upon millions of fish died this way on their way downriver, and indeed only a scant few ever made it all the way to the ocean. Farther to the south, in Sacramento, local officials demanded that residents reduce water usage by 10 or 15 percent to account for declining river flows. Lawns turned to crabgrass, showers got shorter. In the breadbasket of Idaho, the drought caused a more than 30 percent drop in wheat production, cratering yields for the state's most important crop. Farmers harvested two-thirds as many bushels as they would have in a normal year, so little that many of them did not even make back the cost of planting the wheat in the first place. The mayor of one city announced a "day of prayer and fasting for drought relief."

But the consequences of the drought were most profound in central Arizona, where a metroplex of several million people depended on access to water that was now starting to vanish. It wasn't just cattle ranchers like Felkins who were struggling to cope, but also the thousands of cotton and alfalfa farmers who anchored the economy of the agricultural towns outside Phoenix, and the high-dollar real estate developers who needed water access to build new suburbs. More than air-conditioning or the automobile, it had been water that had enabled the existence of this desert metropolis and the carpet of farmland that surrounded it. Pinal County, the agricultural stronghold between Phoenix and Tucson, used more than 300 billion gallons of water a year, more than 800,000 gallons for every resident.

Now, though, the canals and reservoirs that supplied this water were all going dry. The grass was withering and falling apart, leaving behind hostile and lifeless sand. If floods and fires had changed where people lived, the drought was changing where they *could* live, and what they could do with the land they occupied. The era of expansion was over—the cows knew it, and Karen knew it, too.

II.

The New Cotton Kingdom

T he southern half of Arizona sits in the middle of the Sonoran Desert, one of the hottest and driest places on the planet. The area receives between five and seven inches of rain a year on average, or about as much as falls in New Orleans during the month of June, and this rain evaporates almost at once in the scalding hundred-degree weather. Only the hardiest plants—tufty creosote, prickly white bursage, and indestructible prickly pear—can survive on their own in the chalky white silt.

Nevertheless, people have been farming in Arizona for almost as long as they have lived there, which is to say more than four millennia. The available archaeological evidence suggests that even the earliest Indigenous societies were agricultural in nature, centered around the domestication of corn. The Hohokam people, who flourished in the desert during what we think of as the Middle Ages, were even more prolific, growing beans, gourds, squash, and pumpkins.

The Hohokam accomplished this by means of a complex irrigation system: they constructed hundreds of miles of canals across the desert, dug thousands of distinct channels, and designed intricate gates and sluices for controlling water. Every spring, when the winter snowpack melted in the nearby mountains, water rushed down the peaks and poured into the valley along dried-up riverbeds. The Hohokam canal system channeled this water away from the rivers and redirected it toward their villages, where it seeped across the fields. The first Europeans to observe this practice were impressed: a Spanish priest recounted in 1628 that "there are streams of fine water that the Indians employ with no little ingenuity for irrigating their fields."

A couple of centuries later, when settlers from the young United States arrived to prospect for gold and silver, they piggybacked on the old Hohokam canals to build new cities of their own. An ex-Confederate soldier named Jack Swilling founded the town of Pumpkinville in 1867 after observing a ruined irrigation network along the Salt River on his way to a nearby fort. Swilling convinced a bank to loan him money for the Swilling Irrigating Canal Company and set to work refurbishing the Indigenous ditches and locks. The town's location was ideal, but Swilling was a gadabout and a morphine addict, and one of his shareholders soon stepped in to give Pumpkinville a more marketable name—Phoenix. An early map notes that "the town is embowered in shade trees and shrubbery, has streams of living water through every street, is surrounded by orchards, gardens, and vineyards, and is one of the handsomest in the West."

Even so, desert rivers like the Salt and the Gila were capricious and unpredictable, prone to periods of extreme flooding and extreme drought, and this made it difficult for the settlers to expand their agricultural efforts. Under the direction of Teddy Roosevelt, the new federal Bureau of Reclamation undertook a series of monumental water control projects that would "reclaim" the desert land for farming. The government built a suite of dams and reservoirs at astonishing speed—the Coolidge Dam, the Hoover Dam, the Roosevelt Dam, and several others. These projects gave the federal government control over the western water ecosystem: the dams stopped the valleys from flooding during wet years, and the reservoirs collected water to dole out during dry years.

In central Arizona, the largest water control effort was the San Carlos Irrigation Project, anchored by a reservoir near the source of the Gila River.

The original purpose of the project was reparative—the Indigenous tribes along the river had seen much of their river water diverted by the colonial settlers, and the San Carlos project was meant to make agriculture viable on the new Gila River Indian Community reservation. As long as the water was flowing into the valley, though, the government saw fit to let white settlers tap it, too. These settlers used the water to grow cotton, as did their descendants, and the descendants of those descendants.

One of those descendants is named Nancy Caywood. Nancy is a squat woman with a short haircut and a determined walk, prone to putting her hand on her hip. She has spent most of her sixty-five years growing cotton on a set of squarish fields in central Pinal County, using water from the San Carlos reservoir. For decades she has worked through the scorching summers to see each year's harvest to its conclusion, reaping cotton that will be spun into T-shirts, jeans, and underwear. It's work that has allowed her to raise and support a family of children and grandchildren that has spread out across the county.

It is almost certain that Nancy's family will not be able to follow in her footsteps.

Nancy's grandfather was a man named Lewis Storey, who moved from Texas to Pinal County around 1930 and purchased a few dozen acres of land outside a village called Casa Grande. Storey signed a contract with an entity known as the San Carlos Irrigation and Drainage District, which regulated the flow of water through the network of canals that transected the county. He agreed to pay a few dollars a year in exchange for two acre-feet of water from the mountains, or about six hundred thousand gallons. Storey used the water to grow cotton, as did hundreds of other farmers who lived nearby. As long as you had enough water, the ultra-hot climate of Arizona was ideal for agriculture: the sun was always shining, the temperature never dipped below freezing, and there were none of the pesky rainstorms that flooded out young plants or helped weeds spring up in their midst.

Nancy was born in the heyday of the cotton industry, and she remembers

Casa Grande as a small village surrounded by endless fields. The downtown area was always abuzz with farmers and farmhands coming into town to hone their equipment, refuel their tractors with diesel, or discuss loan terms with the agricultural bankers whose offices lined the main thoroughfare. There was a large granary across the train tracks and a massive press that made cottonseed oil, so pungent you could smell it from half a mile away. Down the street from Nancy's house were a pair of enormous cotton gins, each one a hundred feet tall, and when the sun set in the west it turned the gins into hulking silhouettes.

The state of Arizona produced less cotton than more fertile states like Alabama and North Carolina, but the industry was just as essential to Pinal County's economy as it was to any small town in the Old South. Harvest time was a frenzy of noise: there was the determined rumble of the threshing machine, the buzz of planes passing overhead to spray down growth-regulating chemicals, and the whine of the trucks as they arrived to cart away thousands upon thousands of cotton bales. It was hard work, but the Caywood family took pride in being able to raise their own living from the earth.

The burgeoning cotton industry soon expanded beyond the capacity of the San Carlos reservoir. Farmers were flocking to the area faster than the government could build canals for them, and the older families like Nancy's had already claimed all the reservoir water. Fortunately, water had become available from another source—the earth itself. The porous land underneath the desert acted like a sponge for many millennia of river runoff, storing the water in large aquifer beds of sand and stone. Engineering advances had made it possible to squeeze the water up to the surface, allowing for a gold rush of so-called groundwater from subterranean wells. By the time Nancy had come of age, wells provided as much as half the state's water supply and around two-thirds of its agricultural water supply.

But soon the earth, too, proved an insufficient supplier for the cotton boom. As the farmers sucked water out of the ground, the water table underneath them began to sag and sink, creating massive fissures that threatened to split apart streets or swallow homes whole. This was the inevitable result of depleting the aquifers faster than they could regenerate. In 1980, over the furious objections of developers and farmers, the state legislature passed a law that clamped down on new groundwater usage, placing Phoenix and Pinal County on what amounted to probation. The regulations barred farmers from drilling new agricultural wells and mandated that legacy farmers cut down on their usage in order to stabilize the aquifer. Nancy watched as the growing cotton industry seemed to freeze in place around her. Fear set in among the other cotton producers in the county.

Yet again, the federal government stepped in to engineer a solution. This time the necessary water came not from the Gila, or from underground, but from the Colorado River, which flowed along the western border of the state. The snakelike river traversed seven different states on its way from the central Rockies down to the Gulf of California, and these states had been fighting over the river in court for decades, suing and countersuing one another in an effort to hash out who could use its water and for how long. After decades of legal turmoil, the courts had established a system that awarded each state a certain amount of Colorado River water per year, with the excess to be stored in two massive reservoirs until it became needed. The purpose of the allocation system was to prevent "upper basin" states like Colorado from sucking up all the water before it could flow south to "lower basin" states like Arizona, and vice versa.

Arizona's rights were the most junior of any state, but the state had the right to draw about three million acre-feet of water from the river per year. The problem was that the river was hundreds of miles away from Phoenix. In order to use the Colorado River to feed the cotton fields of Pinal County, the state first had to find a way to bring the water to the fields.

After extensive negotiations, the federal government agreed to build a three-hundred-mile canal that could connect the Colorado River with the state's population centers. The so-called Central Arizona Project was one of the great miracles of modern engineering, delivering hundreds of billions of gallons of water a year not only to Phoenix and Tucson but to the vast expanses of farmland in Pinal County, where cotton and alfalfa farmers were desperate for more water. The Colorado provided more water than Arizona knew what to do with, so the state offered the river water to farmers at discount rates, giving them extra incentive to turn off their wells.

It was a win-win: the state solidified its claim to the much-desired Colorado River, and for the first time in decades, the farmers of Pinal County had all the water they needed.

—

In the years before the canal, farming in the county had been restricted to legacy families like Nancy Caywood's. The arrival of the Central Arizona Project opened the cotton farming business to a new generation.

One member of that generation was Don England, a lifelong resident of Pinal County whose distant ancestors had been some of the first farmers to settle the area. Don's father was a refugee from the Dust Bowl and farmed on a small piece of land next to their house when Don was a child, but he died before Don could work the land with him as an adult, and before the federal government built the Colorado River canal. Don came of age wanting to keep his father's memory alive, but he couldn't just break some land and start farming—there the water from the San Carlos reservoir was already spoken for, and the state was cracking down on new groundwater wells. After spending a few years in the military, he went to work as a foreman on the Ak-Chin Indian Reservation, where there was an agriculture operation with ample water thanks to a tribal court settlement. Don relished the opportunity to work on a prosperous farming enterprise with state-of-the-art equipment and a license to experiment—there were even peach orchards on the reservation, something he had never seen before in Arizona.

When the Central Arizona Project canal arrived in the late 1980s, it allowed Don to strike out on his own, and replicate this robust enterprise on his own land. He leased a few hundred acres near the town of Eloy, an area that consisted of nothing but crop fields, and he set to work growing cotton and wheat there. Over the following years he leased more land on various plots until he had more than five thousand acres in and around Eloy. The previous owners of the land had fed their crops using a groundwater well, but that was no longer necessary. As the federal government built the main Central Arizona Project canal, a group of Eloy farmers had banded together to form an irrigation district with a network of smaller canals through the area, extending a spiderweb of concrete ditches from field to field. Don's plot was next to one of those ditches, which allowed him to flood-irrigate a thousand acres of cotton in the span of a few hours.

Farming was hard work, but it was impossible for Don not to feel a rush of satisfaction when he saw the water arrive in the canal. Every few weeks from March until November, he and his team of ditch riders drove down to the fields and waited for the water to arrive. Each field in the county had an appointed watering time, often in the middle of the night or at the crack of dawn, and a farmer had to open the metal gates in his ditches at just the right moment. Don always heard the water before he saw it, rushing down the concrete laterals nearby until it sailed along the length of his cotton field. Leaking through the row of metal gateways, the river water trickled down the ever-so-slight slope of the field, turning the field for a few moments into what looked like a large, squarish bog. After a few minutes, the water sank into the soil, and a few days later the stalks of cotton sprang up.

All around Don's field, bordering it on every side, was more cotton, hundreds of thousands of acres of it. It was almost an ocean of plant matter, so complete and uninterrupted that the earth itself seemed to rise and change color as harvest time approached. To the pilots of the planes that passed overhead to spray growth-regulating chemicals onto the fields, the landscape looked like one big rippling quilt, each square of which sustained a long-suffering farmer who now had the water he needed to survive. There was as much cotton production in Pinal County in the late 1980s as there is in the entire state today, and all of it was made possible by the Central Arizona Project, and by man's newfound control over the water cycle.

Before the canal, Don had debated whether it was worth it to pass his farming knowledge on to his progeny, but the arrival of the Central Arizona Project seemed to suggest that there was reason to keep going, that farming had a future in Pinal County. He even had a worthy successor. His granddaughter Cassy had grown up visiting him at the peach orchards on the Ak-Chin reservation and as a teenager had loved to ride around with him on his tractor as he toured the cotton operation down in Eloy.

Cassy was the fifth generation of England's to grow up around cotton, and she married another cotton farmer from the area, but there had long been a taboo in the England family against women doing the physical work of farming. Don let Cassy's brothers drive the tractor and operate the farming equipment when they were kids, but he never invited her to do the same, and she never felt like she could ask to join in. Even when she worked for him part-time during college, he always restricted her to managing the books, preferring to let her uncle take charge of the farming operation itself. This accounting work had given Cassy an intimate familiarity with the grim realities of modern agriculture. The price of cotton tended to swing up and down from year to year, and any farmer who didn't lock in his crop with a forward contract risked losing all his money if the price cratered around harvest time. Meanwhile, the price of a cotton-picking machine had tripled over the course of twenty years, and labor had gotten more expensive, too. It wasn't clear to Cassy how Don managed to make the math work.

Indeed, several bad harvests decimated the ranks of Pinal County farmers in the early 2010s. Many older families sold off their land to speculators and retired, shoring up their finances and freeing their families from the caprices of the cotton commodity market. It was around this time that Don agreed to teach Cassy how to grow cotton—teach her every part of the process, too, not just the theoretical and financial parts. The family's survival still depended on the success of the larger leased field down in Eloy, so Don couldn't risk training Cassy down there, but there was a hundred-acre plot next to Don's house that had lain fallow for a few years. In the spring of 2013, right around Cassy's thirtieth birthday, the two of them started working that plot of land. Don explained the process in rigorous detail: how to level the fields, how to plant the seeds in the earth, how to till the rows between the seeds, how to open and close the ditches, how to tell when the leaves were healthy, and when to expect the small white tufts of cotton to poke out.

It was a lot for Cassy to absorb, and she sometimes worried she was losing track of it all, but her grandfather reassured her, and thereby himself. She was a quick learner, he said, and the family farm was in capable hands. The road ahead might be rough, but the future was secure. As long as snow kept melting in the Colorado mountains, and as long as water kept flowing through the canal, the cotton would still grow.

III.

Turning Off the Faucets

In May of 2021, Nancy Caywood got in her truck and drove east from Casa Grande, heading for the mountains. On her way out of town, she passed hundreds of cotton fields, some tilled and ready for cotton, others brown and abandoned. Nancy had been accustomed all her life to seeing the land around her explode with green as spring gave way to summer, but this year was different. There was no activity in the fields, and a suspense seemed to hang in the air. Just like Karen Felkins, Nancy was struggling to adjust to a new reality. The rains had not come.

Nancy was driving up into the mountains so she could look at the San Carlos reservoir, the source of the irrigation system that had watered her family's fields for three generations. When she arrived at the reservoir, a serpentine mountain lake a hundred miles away from her farm, she witnessed something her grandfather Lewis Storey would never have believed. The reservoir was empty. In the section nearest the dam, the water level had fallen low enough to expose the striated walls of the canyon, revealing dense clusters of lines like the cross section of a tree trunk.

In previous years, as the megadrought had descended over the West, Nancy had grown used to receiving less than the two acre-feet of water that her grandfather had received. Still, there had always been *some* amount of water flowing through the system each year, even if it didn't arrive until after the summer rains. That was no longer the case. In January, Nancy had received a letter from the San Carlos irrigation district that promised her a meager one-third of an acre-foot, but the district had sent her another letter in March that said she would not be receiving any water after all. The drought had come to a crescendo, and a relentless heat had melted all the snowpack in the mountains, which meant no water had flowed into the reservoir. The lake only had a few thousand acre-feet of water left, and even that was expected to dry up before high cotton season arrived. By the time Nancy drove up into the mountains, the reservoir had dwindled to around fifty acre-feet, so low that state fire crews couldn't even draw on it to fight nearby wildfires.

Even through the rainless winter, Nancy had retained some small measure of optimism, but now doom was staring her straight in the face. She would not be able to plant cotton that year. She didn't have a working groundwater well on her farm, and she didn't have any rights to Colorado River water, which meant she would have no option but to leave her land fallow. Without water, the dirt canals behind her plot would soon dry up and sprout weeds.

The farmers around Nancy all relied on San Carlos water as well, and as the summer wore on most of their fields turned chalky and brown. The lucky few who had wells could still plant a few acres, but everyone else had to watch as their land fell apart before their eyes. Some of Nancy's neighbors had received reduced deliveries before, in the years when the reservoir was lowest, but never had the whole district seen its water disappear. Nancy had always loved the feeling of driving on the rural highways that shot through the farm grid, surrounded by sturdy cotton and waving alfalfa, but now the drive was desolate, horrifying—there was nothing but dirt. The dry soil in the fields needed the irrigated water to hold it together, and in the absence of that water it started to disintegrate in the desert wind, whirling into dust eddies and blurring out the sun. When Nancy went to the mailbox and looked out at the land around her, she cried. She didn't know how much longer she could hold on to a farm that didn't grow anything. And if the drought didn't let up, how could she expect her grandchildren to follow in her footsteps as she had followed in her grandfather's?

Elsewhere in the county, the consequences of the drought were even more profound.

As devastating as Nancy's situation was, it was only temporary: if the drought ended and the San Carlos reservoir filled up, she would receive water again. The farmers who depended on the Central Arizona Project, though, were losing not just their water but also their water rights. The Colorado River was far from dry, but farmers like Cassy England no longer had the authority to use it.

If you visited Cassy's fields in the summer of 2021, you might never have known there was a drought. The fields still roared with green, and the harvest was going off without a hitch, thanks to water from the Colorado River hundreds of miles away. But this water had always come with a caveat. The federal government had financed the construction of the Central

Arizona Project canal out of its own coffers, but the project wasn't free: the government expected to get its money back. The local irrigation district that carried Colorado River water to Cassy's farm was obligated to reimburse the government for a portion of the cost of the canal, which it had intended to do by collecting annual fees from farmers like her. But when the government triggered the repayment obligation in 1993, the irrigation district came up short—it was not collecting enough money to pay the installments on its debt. In 2004, after a decade of wrangling with the government, the districts came to a compromise: the government would grant them debt relief, and in return they would accept weaker rights to the water from the Colorado. Farmers in Pinal County would receive a portion of "excess water" from the river until 2030, at which point their rights would expire. Arizona already had the most junior rights out of the seven states that took water from the river, so the new settlement put farmers like Cassy England at the very bottom of the totem pole. The river supplied water to forty million people in seven states, and they were last in line.

At the time of the settlement, farmers like Don England didn't dwell too much on the expiration date. The year 2030 was a long way off, and a lot could change in a quarter-century. Maybe by that time the state would have found new water from somewhere else or relaxed the restrictions on groundwater so that farmers could pump from underground again. Cotton required a lot of water, but the farmers had found ways to make the process more efficient over time, and many were optimistic they'd be able to survive the settlement.

As it turned out, the deadline arrived a decade earlier than expected. The same drought that drained the San Carlos reservoir on Nancy's side of the county also reduced water levels in the Colorado River, drying up the massive reservoirs that anchored the waterway. The river was nowhere near empty, but there were tens of millions of people who relied on it, and the federal government had to find a way to cut down on usage, fast. In August of 2021, the US Bureau of Reclamation declared a water shortage on the

Colorado for the first time in US history, slashing the amount of water Arizona could take from the reservoirs.

Farmers like Cassy were first in line to absorb these cuts: Farmers in Pinal County would see their water deliveries reduced by a quarter in 2022, once the shortage declaration took effect, and even more in the following years. The leased land down in Eloy could fit around five thousand acres of cotton, but Cassy would only receive enough water to raise around two thousand acres during the 2022 growing season. The upfront costs of harvesting the cotton would stay the same, but she would lose thousands of dollars in revenue from the reduced acreage.

Cassy was optimistic by nature, like many farmers, and she was not prone to desperate displays of emotion, but even she had to admit that the future did not look good. If a one-year water reduction was the end of the story, she might have been able to make it work, but the reduction would be just as large in years to come, if not larger. The old groundwater wells near the Eloy land were out of commission, so it was Colorado River water or nothing. If she kept sizing down her acreage, sooner or later she would have to hire fewer farmhands, lay off her irrigation contractor, or sell off some of her equipment. Plus, she wasn't sure whether the landowner in Eloy would renew her family's lease on the land—there was more residential development in the area than ever, and a lot of land was owned by out-of-state investors who might not want to take a chance on the tenacity of farmers like her. Even if the farmers somehow found a way to work without a quarter of their water, they wouldn't be able to find a way to work without half of their water the following year, and every year after that. Fields were going to go fallow, production was going to fall, and an already impossible business was going to get even harder. The last remaining farm families in the county would soon be heading for the exits, and the rural communities they left behind would shrivel in their wake, forcing more people toward the cities.

Don never lived to see Cassy fallow the family's fields: he passed away just a few months before the 2022 growing season began. In the months after his death, Cassy watched the once-vibrant expanse of cotton turn into a checkerboard, one out of every three or four fields warping with brown even as the rest of them bloomed green. Across the county, the land was drying up, stiffening like a corpse. And corpses attract vultures.

IV.

The Man from Buckhead

The town hall of Florence, Arizona, is a low-slung building in a sanitized pueblo style, unremarkable in the way public buildings tend to be. It sits at the north edge of a small, dusty town, across the street from a library, a few miles west of Karen Felkins's ranch. No one would mistake it for a place of any great consequence.

At six o'clock in the evening on July 1, 2002, a man named W. Harrison Merrill entered this building with the intent to change history. He hailed from the ritzy Buckhead neighborhood of Atlanta and for more than two decades had made a living as the developer of several large planned communities. He had an appetite for risk and an eye for valuable land, and he liked to think big when it came to building subdivisions—not just rinky-dink streetscapes with pleasant-sounding names, but whole self-contained communities, cities within cities.

Merrill had come to Florence that day to pitch the town's council on a development called Merrill Ranch. The plans he described were ambitious: Merrill imagined a community that would someday encompass more than thirty thousand homes, several times as many as existed in the town of Florence itself. The development's first phase would be starter homes for young couples, so that the homes could be "ready for them when employment moves down here"; later there would be higher-dollar homes, "executive housing," plus an entire section that would be exclusive to seniors and retirees. Merrill's company had already acquired about two hundred acres on the outskirts of town, but he planned to acquire thousands more, and he wanted to know if Florence would annex that acreage into the town.

The town leaders agreed—Merrill's plan would bring Florence a king's ransom of tax revenue. A few months later, Merrill's company submitted a two-hundred-page planning document that laid out his vision for Merrill Ranch. The plans imagined "a community designed to capture the rural lifestyle and beauty of the northern Sonoran Desert," extending across more than seven thousand acres and encompassing an entire business district as well as a network of parks and trails. Merrill Ranch would take more than thirty years to build, but in due time the community would swell to around fifty thousand residents and "generate" more than twenty thousand public school students. The subsection on water provision, however, was just a few lines long: taking an agnostic tone, the plans noted only that "there are various methods the developer may use to service the project . . . the manner in which [*sic*] will be determined by the developer at a future date."

Merrill had chosen Florence on the assumption that one day the sprawl of suburban Phoenix would expand to meet the town. As Phoenix had doubled in size, spillover development had come to a chain of farming communities on the east side of Sun Valley. Real estate moguls had bought up land in Chandler, Gilbert, and Queen Creek, sketching out huge planned communities and rushing to build them while the housing market was hot. Within a few years of that fateful town meeting, Merrill's audacious plans for Florence had started to sound more like prophecies. Town officials warned residents to anticipate a "tidal wave" of new construction, and the town newspaper bore witness to the "belches of machinery" that were "the first signs" of the coming community.

Merrill was the latest in a long line of developers who had seen in Arizona the blank canvas for their visions of real estate dominance. The twentieth-century history of Arizona was shaped by the almost total hegemony of developers and home builders—the state boasted the nation's first private retirement community and several other self-contained metropolises. It also had a notorious history of land fraud, much of it sponsored by organized crime groups. Sham developers placed advertisements in the local newspapers of northeast cities promising older couples a plot of land out in the desert where they could retire. All you had to do was call the number in the ad, provide a deposit, sign some paperwork, and the plot was yours. Victims would buy their land, fly out to Arizona to see it, and discover that it was worthless, miles away from the nearest water.

"We're always having a little old couple from someplace up north walk in and say they bought a lot in one of those desert developments and now they want to build on it but can't find it," one county planner told a reporter in the early 1970s, by which time the fraud had become a multimillion-dollar enterprise. "We tell them if they have a helicopter, they can probably get to it. If not, they'll just have to hike in. But they'll have to bring in plenty of water." An investigative reporter for the *Arizona Republic* who helped expose this land fraud was later killed by a car bomb.

In an effort to crack down on fraud, the state in the mid-1970s passed a law that required developers in Phoenix and Tucson to demonstrate an "assured water supply," or to prove that they could provide a hundred years' worth of water to any subdivision they wanted to build. For as long as developers could pump unlimited groundwater, they had no trouble finding that supply, but just a few years later the state passed sweeping restrictions on groundwater usage as well, placing a moratorium on new pumping activity. The Phoenix and Tucson metro areas were growing so fast that residential water usage on its own had started to threaten aquifer stability, even though homes used much less water than farms. Over the following decade, the state took several further steps to restrict new pumping, and the real estate industry fought them at every turn.

The two sides reached a compromise in the mid-1990s with the creation of a novel authority known as the Central Arizona Groundwater Replenishment District (CAGRD). The idea was to allow residential developers to conduct groundwater pumping and "pay back" the groundwater they pumped with surface water from the Colorado River. This would allow developers to skirt around the state's restrictions on subterranean pumping, something farmers in Pinal County had never been allowed to do.

In the case of Merrill Ranch, the CAGRD worked like this. A few years before Harrison Merrill came to Florence, the town's water utility had enrolled in the authority. The utility requested permission to pump enough groundwater for tens of thousands of hypothetical homes, none of which existed at the time; CAGRD approved the request and thus assumed the responsibility for "paying back" that groundwater someday. If the Florence water utility pumped two thousand acre-feet of water in a year, CAGRD would purchase two thousand acre-feet of water from the Colorado River to make up for it. This water could be pumped back into the aquifer or stored in reservoirs until it was needed—the important thing was that there had been no net decrease in the amount of water that was moving through greater Phoenix. When the Florence town council cleared Merrill's development plans, they allotted him a portion of the utility's hypothetical groundwater entitlement. Merrill Ranch could suck the aquifer for all it was worth, and CAGRD would pick up the tab.

If you looked at CAGRD in a charitable light, you could say that the authority acted as a balance sheet for the region's total water flow. To look at it another way, the program was more like a giant loophole in the state's groundwater restrictions, one that opened the gates to rampant development and deferred a difficult reckoning to someday in the distant future.

It took a while for the consequences of this loophole to become apparent, though, as external forces soon stopped Merrill in his tracks. In late 2006, just as Merrill's builders were laying the streets for Merrill Ranch's first phase, the American housing market began to tremble, and a year later it entered a state of free fall. Unable to pay back his loans, Merrill had no choice but to hand the development to his creditors and watch as they sold it for a fraction of what it was worth. The Georgia bank that had backed him later collapsed.

Construction at the Merrill Ranch site itself slowed to a standstill. The first two sections of the neighborhood had already been built, and within these self-contained areas the neighborhood rollout went more or less as Merrill had hoped. The completed homes all found willing buyers, albeit at worse-than-expected prices—snowbirds from the Midwest and Canada, mainly, as well as young families willing to take a gamble on a long commute. The thousand-odd residents of the new neighborhood had everything they required to survive, including a Safeway and an elementary school, but not much beyond that.

Merrill's prophecy had broken off in midsentence, along with those of countless other developers who had marched out from Phoenix in every direction. But no one was questioning the feasibility of massive desert communities like Merrill Ranch, and no one had given up on building in Arizona. The force that had stopped the developers in their tracks was not drought or climate change but a temporary downturn in the market. And what goes down must come up.

V.

Closing the Loophole

Harrison Merrill was not the only developer who had taken advantage of the CAGRD loophole during the housing boom. There were hundreds of hypothetical subdivisions and neighborhoods enrolled in the program across the state. Almost none of these projects had broken ground when the housing market started to spiral, and for a long duration of the recession almost all of them remained unbuilt. Since all future growth in the state depended on water rights, though, these hypothetical developments had the privilege of potential existence: the fact that they had not been built mattered less than the fact that they *could* be built. They were not dead, but dormant.

A decade after the housing market crash, when the wheel of the global economy had spun all the way around, the sleeping giant awoke. Home prices slithered back up, banks opened their checkbooks again, and the growth spurt around Phoenix regained its former pace. The trustees who had bought the Merrill Ranch development at auction now decided that the time was ripe to fulfill Merrill's plans. The bulldozers revved up, the shovels again struck the earth, and streets started to flick out across the desert once more. The same thing happened on other empty lots around the state as developers activated the water rights they had been sitting on for the duration of the downturn.

By the summer of 2021, as Nancy Caywood tore up her fields, Merrill Ranch had begun to look the way Merrill himself had envisioned it. The neighborhood seems to rise like a mirage out of the empty desert that surrounds it, expanding over the course of a few seconds until it hogs the entire span of the horizon. The tile-and-stucco homes within these quadrants are so similar that they almost seem to blur together as you pass them by, the borders of the roofs flowing together. The community is only one-third built, but already it is home to several thousand people, enough to support a hospital and fill the original elementary school. An auxiliary neighborhood is under construction on the other side of the highway, and surveyors are already sketching out a massive thousand-home expansion to the east.

The individual home designs in Merrill Ranch represent a concession to Arizona's looming drought—the front lawns are landscaped with desert rock, the backyards checkered with faux turf. Elsewhere, the development flaunts its water access. The entrance to the neighborhood is framed by two massive water fountains that toss out rippling sheets of water beneath gilded serif signs, while the central boulevard runs past a large lagoon dyed with faux-blue coloring and surrounded by genuine grass. In the model home complex on the family side, a sign promises not only a "lap/diving pool," a "tot pool," and a "beach entry leisure pool," but also a "big splash water park" complete with misty spray showers. Meanwhile, on the other side of the neighborhood, in the "active adult" section, the homes look out onto the gentle slopes of an eighteen-hole golf course. Harrison Merrill no longer has any financial stake in the development, but like the great king Ozymandias, his works have outlasted his own involvement.

There are dozens of other such oases throughout the valley, the long-delayed pet projects of other builders and visionaries, stacked with fishing ponds, golf courses, and pickleball courts. It was not inevitable that any of these communities would ever be built, or that people would descend on them to get jobs, marry, start families, and retire. This daisy chain of sun-baked sub-urbs only exists because of the loophole opened by CAGRD, and its contin-ued existence represents a water debt that future generations will need to meet.

As it turned out, the federal government's water shortage declaration in the summer of 2021 was only the beginning of the Colorado River cri-sis. The next year, the federal government announced that the river had reached another shortage level, falling into so-called Tier 2 conditions, and projections showed that in future years it was only going to get drier. The new shortage declaration meant that Arizona would absorb further cuts to its Colorado water allocation, and these cuts would extend far beyond the Pinal County farming community. The second round of cuts would take away water from a wide assortment of suburbs, independent cities, and industrial firms around the Phoenix area. The parties who lost this water would need to go back to groundwater, buy water from other parties, or find a way to constrain their water consumption.

In the years before the drought, a developer who wanted water rights could buy out local farmers and redeploy the water rights attached to their farms in service of new development. Nancy Caywood and Cassy England had watched several of their farming neighbors sell out to developers in this way and use the money to retire. Now that the farmers in Pinal County had lost their rights to water from the Colorado River, though, developers had to look elsewhere to sustain their new subdivisions. For most of them, the sole remaining option was to enroll in CAGRD so they could pump groundwater, pushing their new subdivisions through the loophole.

The problem was that CAGRD had its hands full with the task of finding water for the subdivisions that were already enrolled. Hundreds of dormant developments had claimed their water rights, ratcheting up the pressure on CAGRD to buy more water even as the main sources of that water dried up. The authority relied for decades on its ability to purchase excess water stored in the Colorado River reservoirs, but the drought threatened to make such excess water nonexistent. The Tier 2 shortage on the Colorado forced cuts to residential water use as well as agricultural, which deprived CAGRD of the water it used to "pay back" new groundwater pumping. Furthermore, the cost of water skyrocketed in tandem with demand, making membership far more expensive: the annual CAGRD fee for households in Pinal County almost quadrupled between 2000 and 2015, rising from $160 a year to $633 a year. Nevertheless, the program's leaders have hesitated to consider the effect of drought on their water purchases, with the agency's director making the perplexing comment that "evaluation of shortage impacts on the Plan is outside the scope of the Director's review of the Plan."

It is unlikely that CAGRD will fail to fulfill its obligations to existing developments like Merrill Ranch, but there may come a time when the program can no longer find any new water and thus must cease to enroll new lands. If that happens, it would place a ceiling on growth in Pinal County, barricading Phoenix's endless sprawl.

Even assuming CAGRD can find an infinite supply of makeup water, there is no guarantee that developers can pump groundwater forever. The deeper you dig a pump, the less certain you can be about the quality of the water you're pumping, and it has become common in Pinal County to pump from depths exceeding one thousand feet. A community like Merrill Ranch may show up as a zero on the state's balance thanks to CAGRD, but that doesn't mean it will always be able to rely on an adequate supply of water. Furthermore, even despite all the state's efforts to manage aquifer depletion, land subsidence in Pinal County is far from over. In 2017, for instance, surveyors discovered a two-mile fissure just outside the county's largest array of cotton fields. The fissure was ten feet wide and more than thirty feet deep in some places.

In the summer of 2021, the state placed a moratorium on new developments in Pinal County, responding to a study that found groundwater demand was on track to outpace supply by around eight million acre-feet. Officials from the state water department told Pinal County officials that developers who wanted to build in the county could not use groundwater to demonstrate that they had water access, closing the CAGRD loophole; given that the Colorado River was tapped out as well, the decision left developers without an obvious alternative. The consequences of the announcement were difficult to parse, but it seemed to many observers that residential growth, much like agriculture, now had an expiration date.

"I don't want everybody to panic," said one county commissioner after the announcement. "If you have a house in Casa Grande today, you will not run out of water. If you are building a house today you will be fine. We are working on solutions for water issues that will come into play decades down the road, not tomorrow."

The comment was revealing. The threat that loomed over Pinal County was not that existing residents would run out of water, but that the pace of development will grind to a halt, ending a speculative boom that has lasted for the better part of a century. The families who move to an isolated area like Merrill Ranch are making a tacit bet that one day a full community will spring up around them. If the outward growth of the Phoenix region stops in its tracks, then the people on the farthest outskirts will be left in the lurch, stuck in the desert with no promise of future growth around them.

The price of a home signals not only its present value to the people who live in it but also its future value to those who might buy it down the road.

If the state restricts development on the outskirts of Phoenix, it jeopardizes the property values that have been the anchor of middle-class prosperity. The aftershocks of such a decision would be profound. Having sprawled out as far as it could, Phoenix would have to stop in its tracks and roll backward, reversing its imperial conquest of the desert. This kind of climate displacement is not as straightforward as the kind that happens when a hurricane destroys a neighborhood, but it, too, will alter the balance of who lives where in the United States. Instead of forcing people who already live in Pinal County to move elsewhere, water insecurity will drive away those who might have chosen to move there in the future.

Water is essential to life and to growth, to the insatiable demands of capital and profit. Until now, the farmers and ranchers have borne the cross of the western megadrought, but before long the shortage will come for residential water as well. Nobody's faucet is running dry yet, but for the next generation of Harrison Merrills, and the high-rolling businessmen who finance them, the imperative is clear: find more water, or go somewhere else.

VI.

Where the Lemon Trees Grow

Robert Stone wheels his white SUV away from the main thoroughfare and onto the unpaved farm road, gesturing out the window at field after field of lush green alfalfa. There are tractors and trucks buzzing back and forth along the length of these fields, helping to mow down the crop and roll it up into big bales. Because alfalfa is a year-round crop, the fields are all in different stages of growth: some are mere patches of stubble, others are rippling and mature.

It takes Robert almost ten minutes to traverse the length of the alfalfa complex, at which point he turns left and drives past a long grove of olive trees; at the far edge of the grove is the staging area for a mobile olive oil press, which shows up during harvest time so that the farmers can press their fresh olives into oil without having to send them to an off-site plant. In the distance, beyond the grove, Robert points out to me the previous site of an experimental field for growing oranges and watermelons. There's cotton, too, all of it growing tall and vivid green.

But it's the lemon grove that stands out the most, maybe because of how it looks from the farm road, framed against the dry slopes of the drought-stricken mountains behind it. We drive past several dozen rows of leafy lemon trees, so tall and bushy that their branches are starting to get tangled up with each other. The ground underneath them is dry, the same pale green as barnyard hay, but the lemon trees themselves are about as healthy as trees can be.

The farming enterprise where Robert and I are driving is a mere fifteen miles away from the brown fields around Nancy Caywood's house, but it seems to exist in an alternate universe. There is no water shortage here, nor are there any restrictions on who can grow what crops or how much of them they can grow.

Despite this farm's geographic proximity to Pinal County, it is separated from the county by a very important jurisdictional boundary. This farm is located on the reservation of the Gila River Indian Community, a federal tribe that controls more water rights than the city of Phoenix. It is no exaggeration to say that Arizona's future will be decided on this farming complex. The water that helps nourish the lush lemon groves could also sustain years of future residential growth outside Phoenix and Tucson, but only if the tribe agrees to sell it.

Robert is a thoughtful man, with wide eyes and a ponytail. He walks like a man who is accustomed to stooping down, and his hands are at once large and agile, capable in equal measure of digging a ditch or of thumbing open a flower.

Robert has worked on the tribal farming enterprise for most of his life: he started out as a tractor hand when he was just a teenager, in the late 1960s, and moved up the ranks over the years to become general manager. Having spent so much time on the farms, he knows just how transformative water can be.

The legal landscape surrounding Native American water is very complex: court decisions have established that federally recognized tribes have expansive and ironclad water rights, but those rights are often difficult for the tribes to claim and realize. The Gila River tribe, for instance, possesses theoretical rights to all the water it used before European settlers arrived, or in other words the entire flow of the Gila River.

Before settlers like Jack Swilling arrived, the Gila had been a mighty waterway, but it was diverted and dried up by the settlers over the course of the early twentieth century, making irrigation impossible and leading to an epidemic of diabetes on the drought-starved reservation. The traces of that epidemic were still visible in obesity and gout when Robert was growing up, as were the lingering consequences of the Gila River's decline. The farming enterprise was just a few hundred acres, and most tribal households relied on faulty wells to get their drinking water. In theory the reservation received irrigated water from the same San Carlos system that served Nancy Caywood, but the water entered the reservation on dirty, unlined canals, and there was never enough of it, since farmers like Caywood diverted it before it could reach the reservation.

In order to claim its water rights, a tribe must sue the federal government and reach either a settlement or a trial judgment. The Gila River tribe filed its lawsuit in 1974, when Robert was still a teenager working on the farming enterprise, but the suit did not reach a resolution until thirty years later, by which time he was preparing to take over the enterprise as general manager. The tribe's initial suit had claimed a million acre-feet of water, equivalent to the old flow of the Gila River, but in the end the tribal leaders agreed to settle with the government for around half that amount, still an astonishing volume. This tribal settlement was part of the same 2004 agreement that established an expiration date for Pinal County agriculture, and indeed the two outcomes are related: some of the water allocated to the Gila tribe in the settlement was taken from the Pinal County farmers, who relinquished it in exchange for debt relief on the Central Arizona Project canal.

To be sure, most Native American tribes were not as well-off as the Gila when it comes to water. In the northern reaches of Arizona, on the Navajo reservation, thousands of people were languishing with no water at all—the tribe had never reached a water settlement, and thus there was no water delivery infrastructure on the reservation whatsoever. In times of drought, the lakes and ponds that provided almost half the tribe with clean water all dried up, and so did the isolated groundwater wells on the reservation, which forced many Navajo families to rely on unreliable deliveries of bottled water. Across the country, around one in ten indigenous Americans lacked access to safe and clean drinking water.

The Gila had suffered under similar circumstances in Robert's childhood, but the settlement had given them more water than they could use. This new water began to arrive on the reservation around the same time as Robert took the reins of the farming enterprise, and he used it to expand the farming operation in ways he had never dared to imagine. He doubled and then tripled the farm's production of cash crops like hay and alfalfa, and revived old experiments with olive and lemon trees. The results weren't always successful, but the farmers in his employ learned something every time, and the farm was bringing in more money than it ever had. There was so much water that the farm sometimes struggled to find a place to put it: they dug and filled a large reservoir near the back of the farm and expanded the canals so they could move more water for longer. As we drove around the farming tracts, I couldn't help but notice it foaming out of spouts and rolling back and gushing down long ditches.

The tribe had so much excess water that its leadership had to invest millions of dollars in underground storage infrastructure that allowed them to stow away the water they couldn't use on the farm. After he showed me around the farm, Robert took me to see one of those storage facilities, a kind of supercharged underground aquifer that lay below the old bed of the Gila River. As we approached the aquifer from across the desert, the SUV was suddenly surrounded by thick tangles of shrubbery, long grasses, and healthy clusters of trees. None of this had been here when Robert was growing up—the river had been a corpse, dry and cracked with age—but pumping the water beneath the riverbed had rejuvenated life on its banks. In this small segment of the river, the tribe had managed to restore the desert environment that had existed before the Gila was overdrawn and racked by drought.

As another condition of the settlement, the Gila River tribe has the legal authority to sell off a portion of its water rights in exchange for cash or infrastructure investment. The sale price of this water is capped by the settlement, but the tribe has quite a lot of it—the settlement awarded them more than a quarter of all Colorado River water that passes through the Central Arizona Project aqueduct, more than enough to make or break the fortunes of any water-starved suburb. Indeed, one of the tribe's first water deals transferred around thirty thousand acre-feet per year to the balance sheet of CAGRD, allowing the authority to replenish the vanishing groundwater reserves being drawn by residential developments like Merrill Ranch. As the Colorado continues to dry up, and as groundwater pumping becomes riskier, it is safe to assume that CAGRD will be darkening the tribe's door again.

As Robert sees it, the tribe is at a crossroads. When he was a child almost every family on the reservation maintained its own personal farm and the farming enterprise was a main source of income, but things are different now. The main driver of the reservation economy these days is the casinos the tribe has erected on the outskirts of Phoenix, and Robert is one of the only people around who still bothers to grow his own okra and tend his own backyard garden.

The revenue from future deals could help improve conditions on a reservation where almost half of all residents live below the poverty line, but selling off water rights could also mean ceding a centuries-old heritage, giving up on a tradition of farming that dates to the earliest inhabitants of the desert. The ultimate beneficiaries of such a decision would be developers and real estate moguls, snowbirds and suburban mayors, descendants of the same people who drained the river in the first place. The future of Phoenix and the real estate growth machine that built it now depend on the tribe's decision. And as is ever the case with climate change, the question seems to be not whether this desert society *can* continue to grow, but whether it *should*. How do we decide when to stop building, and what do we do when there isn't enough to go around? When does a place put up a NO VACANCY sign and tell people to move somewhere else?

CHAPTER SEVEN

Bailout

RISING SEAS AND
FALLING MARKETS

Norfolk, Virginia

I.

The Tremor

S ara Langford and her husband, Phillip, moved to Norfolk, Virginia, in June of 2017, right after they got married. As they arrived in the coastal city, they could see a bright middle-class future unfolding before them: Phillip was starting a promising residency at the local naval hospital, and they figured they would only have to rent for a year before they could afford to buy a house and start a family.

Sara's family hailed from Texas, and the couple had been living in Houston before they moved to Norfolk. A mere two months after Phillip started his residency, Hurricane Harvey struck their old hometown, destroying their previous home; they had sold it only weeks earlier. The storm also flooded Sara's mother's house, and her sister's house, and many of her friends' houses as well, forcing all of them to hole up in hotels for weeks. As Sara watched the news from her apartment in Norfolk, she saw TV anchors sailing through familiar streets in speedboats and helicopters plucking victims from the roofs of houses that looked just like the one she had grown up in. She had gotten out just in time.

A year later, the Langfords started searching for a house in a well-to-do area called Larchmont, situated on the banks of the Elizabeth River. The neighborhood looked like paradise to Sara: young children scratched chalk doodles on the sidewalks, college students and senior citizens ran side by side on the river trails, crepe myrtle trees popped pink along the lengths of whisper-silent streets. The homes were charming, even idiosyncratic, and there was a commercial strip nearby with a few trendy breweries and restaurants. The neighborhood public schools were some of the city's best and many residents had signs on their lawns that proclaimed WE ♥ OUR CHURCH. It was easy for Sara to imagine herself exercising in Larchmont, worshipping in Larchmont, settling down and starting a family in Larchmont.

The Langfords visited three houses in the neighborhood with their realtor, Gabriella Beale. As the couple toured the area, they noticed something about the homes they were seeing. Every single one was in a FEMA-designated flood zone, and as a result they all came with flood insurance premiums that made Sara's jaw drop.

"We were looking at one house close to the water, and Gabriella started talking about flood insurance," she recalled. "I said, really? In this area?" The house was about half a mile from the water, but the monthly flood insurance premium was $850, almost as much as the mortgage payment itself.

Like Texas before Hurricane Harvey, Virginia was a "buyer beware" state, which meant home sellers and agents had no legal obligation to disclose a home's flood history to potential buyers. Many realtors in Larchmont played down the neighborhood's flood risk, wary of scaring away customers, but Beale's conscience didn't allow her to do that. After Sara and Phillip had looked at a few homes in the neighborhood, Beale said there was something she needed to show them. She drove them to a section of the neighborhood they hadn't seen before, a waterfront street called Richmond Crescent. Almost as soon as they turned onto the street, Sara felt something turn in her stomach. Every home along the water had been elevated off the ground. The one-story ranch houses, which Sara could see had once been charming, now looked like a long row of misfits, perched atop giant slabs of concrete. A construction team was in the process of raising one house as the couple drove down the street— an intervention that Beale said would likely cost more than the house itself.

In decades past, Beale told them, flooding had never been an issue in the neighborhood, but as the sea levels around Norfolk had risen, flooding events had become far more common. Now some streets in Larchmont flood at least a dozen times a year, with almost every high tide, and the wrong combination of rain and wind could turn the neighborhood into a labyrinth of impassable lakes and puddles. One main road flooded even when it didn't rain, and pooled up with puddles for days after even a minor drizzle.

For Sara, whose family was still recovering from Harvey, the raised homes were a deal breaker. She knew she would never feel at peace living in an area that was so vulnerable to flooding, not after what she had seen stormwater could do.

"When I saw that I was like, absolutely not," she said. "I said, we're just not even considering the area anymore." The Langfords ended up buying a house two neighborhoods over, in a slightly blander area called Colonial Place. They chose a home that sat just outside the floodplain and didn't require flood insurance.

Just a few months after they moved in, though, they started to find that some blocks in their neighborhood became swamped with water after every heavy rain. In the days after a big storm, the estuary at the western edge of the neighborhood tended to spill over into the lowest-lying streets, cutting off one major thoroughfare and pooling around the tires of parked cars. In autumn, when there was a king tide, salt water sloshed through Colonial Place from the east. Sara and Phillip had tried to mitigate their risk, but the water had found them anyway.

The Langfords' home hasn't yet been damaged by these minor flood events, but they often have to change their dog-walking route because their favorite trail is ankle-deep in water. Their neighbors told them the water problem seems to get worse with every year, and Sara suspects that before long their home will get mapped into a flood zone as well—indeed, given that flood maps are only updated every decade or so, the house may already be in one. By that time, the couple will have moved on to a new navy posting, but they aren't the only ones who have been scared off by the sight of Larchmont's elevated homes.

Beale, who does most of her business in Larchmont, says she's seen several potential buyers decide to look elsewhere after seeing the elevated homes. In one case, she recalled, she had a customer who was sitting at the kitchen table of the house he wanted to buy, preparing to sign the mortgage paperwork. Beale had a pang of conscience as she watched him read the documents and insisted on driving him down Richmond Crescent before he closed the deal. After taking the drive with her, the client backed out.

"There's enough in the news now about flooding that every single buyer has some knowledge," Beale told me. "Some buyers are willing to deal with it," she said, but "some are saying absolutely no flood zones." For the past few years, amid the post-pandemic housing boom, Norfolk has been a seller's market, with most homes getting snapped up in a few weeks, but there's no telling how long that trend will last, or what will happen to home prices in Larchmont if demand starts to dry up.

Another real estate agent put it more succinctly when he referred to the perils of buying a house near the water in Norfolk.

"If the street name ends in 'Crescent,' I would run away," he said, referring to the streets that run alongside the city's rivers. "Either run away, or get a kayak."

There is something going on in Larchmont and neighborhoods like it in other coastal cities. It's a tremor, if you will, or a wobble. As sea-level rise causes more frequent flooding in coastal areas, a mounting anxiety about the impact of climate change has begun to destabilize the real estate market in towns and cities along the Eastern Seaboard. There is no historical precedent for this phenomenon, but it bears some resemblance to the first stages of past asset bubbles, in particular the housing crash that led to the Great Recession.

Long before cities like Norfolk are underwater, the housing markets in these cities will start to reflect the reality of climate change and accelerated sea-level rise. When that happens, the value of vulnerable homes in these areas will begin to decline. No one knows when this change will occur, or what it will look like when it does, but almost all experts agree that at some point coastal property values will fall, and that once they fall they will not go back up.

You can imagine each of the homes in Larchmont as a stick of dynamite with a very long fuse. When modern society began to warm the earth, we lit the fuse. Ever since then, a series of people have tossed the stick of dynamite between them, each owner holding the stick for a while before passing the risk onto the next owner. Each of these owners knows that at some point the dynamite is going to explode, but they can also see that there's a lot of fuse left. But as the fuse keeps burning, each new owner will have a harder time finding someone who will take the stick off their hands.

Norfolk is at an ambiguous point on the length of the fuse. Consider that the payment term for a standard mortgage loan is thirty years, and that the median length of homeownership is thirteen years. Now consider that Norfolk is around five feet above sea level. Climate scientists believe that it could take between seventy and a hundred years for the water in the area to rise five feet. How many more times will the dynamite change hands before it blows up?

The events that took place in Kinston and Santa Rosa represented some of the country's first halting efforts at a managed retreat from vulnerable areas; after floods and fires, public and private entities sought to push people away from the zones of highest risk. The ongoing tremor in Larchmont forebodes something much larger, on the order of the housing slowdown that may arrive in the West as drought intensifies. The pace of sea-level rise threatens not just the survival of one island, neighborhood, or city, but the edifice of the coastal real estate market, home to around 30 percent of the country's population. A downturn in a market like this would jeopardize trillions of dollars that are held by homeowners, insurers, banks, and all levels of government. There would be no winners in such a collapse, but the biggest losers would be the homeowners, the millions of people stuck with homes they would be unable to sell.

When we talk about climate adaptation, the conversation often focuses on how we can safeguard people from suffering and dispossession—how we can protect our communities from wildfires, or how we can help people rebuild after a hurricane destroys their home. The future of a city like Norfolk hinges on far more difficult questions: What should we do with the dynamite? For how long should people be allowed to keep passing it around? Who should be responsible for throwing it away? And how can we move everyone out of the blast radius?

As of right now there are no answers to these questions, and indeed there are very few people willing to even ask them. For the moment, the coastal housing market is a three-way staring contest, with homeowners, governments, and real estate industry figures all straining to keep their eyes open, to keep up the pretense that everything will be fine.

Sooner or later, though, someone is going to blink.

II.

Raise or Raze

Around thirty-five million years ago, during a period when much of the Eastern Seaboard was underwater, a massive meteor about two miles wide crashed into the ocean a few miles north of the Langfords' house in Norfolk. The impact from the meteor left a crater on the ocean floor the size of Rhode Island and three times deeper than the Grand Canyon. When the ocean receded and created the Atlantic coast, the impact crater became what we now know as the Chesapeake Bay. Rivers drained down into the depression and the coastal land itself also started to slide down the slope of the crater, sinking ever so slowly over thousands of years.

In May of 1607, a hundred settlers led by Captain John Smith sailed through the mouth of the Chesapeake Bay and established the Jamestown Colony up the river from Norfolk. The outpost they built on the James River became one of the first English colonial ports on the new continent. The colonists did not know about the crater, but they appreciated that the mouth of the bay offered a large and accessible natural harbor, a great gullet of water that was only twenty miles from open ocean and never froze over.

For more than four centuries the harbor that surrounds Norfolk has been the economic linchpin of the Middle Atlantic. First enslaved people, then cotton, then raw metals flowed through the harbor, and at the turn of the twentieth century the US Navy built its global headquarters along the city's western edge. Norfolk's very existence depends on its proximity to the water, and for most of its history its residents regarded that proximity as something to be cherished. The city's tram system is called the Tide, its minor-league baseball team is called the Norfolk Tides, and its mascot, as seen on signs around town, is the mermaid.

The problem with proximity to the water, though, is that you can drown. The sea level in Norfolk has risen more than a foot since 1950, and it will rise at least another foot over the next twenty years, faster than almost anywhere else in the country. As in Louisiana, this is in part because the land is sinking as the city draws out subterranean groundwater, a process accelerated by natural subsidence along the impact crater. But the Chesapeake Bay also happens to be next to a bend in the Gulf Stream that pushes water up through the Atlantic. As the glaciers melt and scramble ocean temperatures, the current will slow down and "park" in front of Norfolk. The scientific context is complicated, but the result is simple enough—more water, and fast.

This advancing ocean is now on a collision course with the waterfront cities we have built alongside it. Precolonial Virginia was undisturbed marshland, dense with hundreds of turgid creeks and speckled with bug-filled swamps. As the colonial towns around the Chesapeake Bay grew from shipping outposts into port cities and then into industrial centers, they molded and manipulated the natural environment, taking whatever they needed from the water with no notion that the water might someday take something back. A map of the city from just after the Civil War shows that huge portions of the city, from the opera house and the art museum to strip malls and chain restaurants, sit on what used to be creek beds and marshes. In the nineteenth and twentieth centuries, planners began filling in whole creeks and rivers with dredged-up dirt, then building on top of the new land, just as developers would later do in cities like Houston. The decision to subdue these waterways was in part the product of ignorance about how water systems worked, but it was also the inevitable result of local government's pro-development policy: city councils were all too eager to approve new construction if it juiced the tax base, and federal aid money was never hard to come by after a bad storm.

For decades, the consequences of this risky construction were difficult to discern, except when a hurricane or a nor'easter drenched the region in rain. If you looked at the tide gauges, you could see that the sea was coming up by a fraction of an inch each year, but the change was happening too slowly for anyone except the marine scientists to notice. It was not until around thirty years ago, when the expansion of the oceans began to accelerate, that the water level rose high enough to disturb the urban equilibrium.

Like the process of aging, this collision was imperceptible until it wasn't. The first signs appeared about fifteen years ago, when high tides started to eat away local wetlands and erode bulkheads that had once stood sturdy. Groundwater started to seep up through cracks in the asphalt after heavy rains, storm drains filled up faster and stayed clogged for longer. It happened downtown by the art museum and the opera house, in quiet neighborhoods along the city's twin rivers, in bungalow blocks by the beach and shopping centers near the highway. Not even the military could keep the water out: as the ocean rose around the 3,400-acre naval base, nooks and crannies started to flood all the time—the fire station, the electrical system, the telecom facility, the steam pipe system, and the ammunition depots. The routine intrusions of water that city officials called "nuisance flooding" became far more than just a nuisance. Tides submerged cars in salt water, subterranean moisture caused the brick facades of old homes to crumble and disintegrate, and high tides on weekend mornings blocked church parishioners from attending Sunday service. The gradual blurring of the line between land and water, a process that was supposed to take centuries or even millennia, was happening fast enough that you could watch it with your naked eyes.

The rising seas had raised the baseline for flooding in Norfolk, so that residents ceased to see floods as occasional disasters and instead came to see them as ordinary events. This change made daily life in the city not just more dangerous but also more expensive, creating a massive cost burden that had to be distributed through the local housing market: either homeowners had to pay more, or their homes had to become less valuable. Thus the clearest harbingers of climate change in Norfolk were not waves lapping against street corners, but bills arriving in mailboxes.

In the summer of 2019, Alex and Kezi Lane moved to the city of Portsmouth, just across the river from Norfolk, buying their first-ever home on a quiet street in the city's small downtown. Alex was on a Coast Guard career track that required him to do a series of two-year assignments, so the couple had to fly in from Alex's deployment in Puerto Rico to go house-hunting in the area, and they didn't have time to spend more than a few minutes at any single home. The millennial couple visited two dozen properties in a single weekend before settling on one half of a duplex near the water. The house cost just under $200,000—"It was more a matter of practicality than anything else," Alex told me. It wasn't a dream home, but it would be easy enough to sell in a few years when the couple had to move on to Alex's next deployment.

Looking back now, Alex realizes he should have seen the signs. The real estate agent who represented the seller didn't mention anything about flooding in the home or on the street, even though the region was notorious for flooding. A few weeks later, when it came time for the couple to close on the house, the Lanes' loan officer told Alex and Kezi that they would need flood insurance, something that hadn't come up in any previous discussions. Kezi called more than twenty insurers before finding a company that would cover the house. That seemed strange at the time, but it was too late for the couple to back out—their first child had just been born, and they had neither time nor energy to go back to Portsmouth and house-hunt all over again. Kezi got a quote for a flood insurance policy backed by the federal government's National Flood Insurance Program. The plan came with a staggering $3,200 annual premium, but the Lanes didn't have a choice. They bit down and bought the home.

Before they even finished unpacking, they could see something was wrong. The street in front of the house flooded once every few weeks: sometimes the water sloshed up from the river, other times stormwater pooled along the sidewalks, and still other times it gurgled up from the storm drain. It smelled awful, and the couple worried about their son wading into the water by accident when he learned to walk.

In March of 2020, the couple received a letter from FEMA, which administered the National Flood Insurance Program. The letter informed them that FEMA had "determined that your property has experienced repetitive flood losses," and that as a result "this property is now considered a Severe Repetitive loss property." This meant that the house had experienced four or more flood events costing at least $5,000 apiece in the previous two decades. The Lanes had known their home was in a flood zone, but the seller hadn't told them it had flooded in the past, and they hadn't been entitled to see that information before buying—thanks to federal privacy laws, only the owner of a home can request a flood history report from FEMA. A few months later the couple received a renewal notice and learned that their new flood insurance premium would be around $10,700 a year, a more than threefold increase.

In response to the Lanes' entreaties, FEMA sent a claims report showing that the home had flooded at least eight times in the previous twenty-five years—in 1994, 1998, 1999, 2003, 2006 (twice), 2009, and 2011. Alex thought back to his conversations with the previous owner and remembered that the owner had only bought the home in 2014, which meant that the owner may not have known about the home's flood history, either. It had taken the officials at FEMA almost ten years to catch up on their paperwork and reclassify the home as a severe loss property, leaving the Lanes to catch the dynamite just as it blew up.

"The last claim happened nine, ten years ago, and now, four owners later, we're biting the bullet on it," Alex told me. "It cost us thousands and thousands of dollars, and a lot of stress, and maybe the ability to sell our home." By 2021, the new premium strained the Lanes' household budget, especially as they prepared to welcome a second child, and it also threatened to make their home unsalable: Who would want to buy a tiny townhome that came with a $10,000 annual insurance bill? Still, hanging on to the house wasn't an option: the Coast Guard needed Alex to be in Florida by July of that year and there was no way the Lanes could take on two mortgages at once. They needed to find some way of reducing their flood insurance costs before they put the home on the market.

Alex did some research on the internet and decided to call a man named Mike Vernon, who ran a company called Flood Mitigation Hampton Roads. Vernon made a living helping people lower their flood insurance bills, a unique but booming trade in a region where almost everyone lives close to the water. A career salesman with a background in finance, Vernon knew almost nothing about flooding until around 2012, when he started talking with a friend who made a living installing "flood vents," or baseboard-level slats that allowed water to drain out of flooded houses. The friend had been marketing the vents as a feel-good addition for homeowners worried about their stuff getting soaked, but Vernon saw another path: flood insurance costs were going up every year, but FEMA offered a discount for homeowners who undertook "mitigation" measures like vents. He got licensed as an insurance agent and started helping customers figure out how to shave down their insurance policies through interventions like vents, racking up commissions as he went.

Vernon soon realized that the flood risk in many parts of the region was so extreme that vents wouldn't suffice to reduce premiums. He started offering more creative solutions to homeowners in areas like Larchmont: one option was to fill a basement with sand, which helps prevent damage to the rest of the house, but homeowners could also shave their premiums by moving appliances out of a garage or emptying out a sunroom. Sometimes he could save his clients money by finding errors in their policies, but often he found himself forced to tell homeowners that they would need to spend tens of thousands of dollars if they wanted to cut their costs. The problem was not how the homes had been built, but that they had been built at all.

This was the case for Alex and Kezi, whose house demanded something more than mere tinkering. The only solution, Vernon said, was to raise the house above the ground like the ones in Larchmont, a procedure that would cost far more than the home itself was worth. The Lanes did not have the money to pay for an elevation, and plus, their home was a duplex, so the other unit would have to sign up for the elevation as well. If they didn't elevate the home, the premium would only go up.

"We were only twenty-seven, and this was our first house," Alex told me. "We didn't have the money for that. We just didn't."

It is worth pausing to consider what a $10,000 flood premium means. We understand that it's much more than we pay to insure our cars, our homes, or even our lives, and it's easy to see how it might be difficult for a family to afford. But a premium is also trying to communicate something to the person who gets the bill. It's trying to tell the person that she should not be living where she does, that no public or private system can tolerate the risk she continues to incur by remaining in her home. If she cannot afford the cost of her insurance, the premium says, then neither can the insurer.

The central premise of flood insurance and insurance in general is that disasters are rare events: a home might have bad luck and flood several times, but it won't *always* be underwater. Sea-level rise has scrambled that calculus. The flood risk in a riverine city like Houston may grow by degrees over time as hurricanes become more powerful, but that risk will never rise to 100 percent. The flooding in Norfolk, on the other hand, is inevitable, and the implications for local housing markets are far more profound. As the writer Brooke Jarvis put it in the *New York Times Magazine*, the sea-level rise in the area has turned risk into certainty.

As the water level keeps rising, thousands of homeowners like the Lanes will have to make a choice that Vernon describes as "raise or raze." If someone owns a home with an albatross premium like the one Alex and Kezi had, they will never be able to sell that home at a decent price, especially not if the housing market is already weak. A homeowner's options are either to spend an enormous amount of money to raise the home off the ground, thus lowering the premium, or to cut their losses and walk away.

For a cash-strapped millennial household like the Lanes', home elevation was never an option. The house wasn't worth much to begin with, and the couple needed to dump it as fast as they could. Alex spent months talking on and off with Vernon, to no avail; Vernon had tried every angle, courting private insurers and examining obscure elevation documents, but there was nothing he could do to get the premium down. The Lanes put the house on the market, hoping a cash buyer would take it off their hands and flip it, but the listing expired; they listed it again, and it expired again. Halfway through the couple's consultation with Vernon, FEMA wrote the Lanes again to inform them that under the agency's new system for calculating flood risk, their premium would now be $13,000.

At first Vernon suggested that the couple consider filing for bankruptcy and letting their bank repossess the home. Their credit score would take a hit and they would have to wait several years before they stood a chance of getting another mortgage, but at least the house would be out of their hands. When the couple ruled that out, Vernon went back to work, and after months of calculation and comparison he convinced the county to issue the Lanes a new elevation certificate that showed their home was a little higher off the ground than the county had said it was. The couple also spent several thousand more dollars to fill the basement with sand on Vernon's recommendation. The new certificate and the filled basement brought the home's FEMA premium down to $4,300 a year—still high enough to be an albatross on its market value, but not high enough to make it unsalable. If the house had been damaged by a hurricane, the city of Norfolk might have bought it out with funds from FEMA, or at least FEMA could have given the Lanes some aid money, but there was no federal funding available for *potential* flooding, even if the threat of future damage had almost been enough to drive Alex and Kezi into bankruptcy.

The Lanes put the house on the market in May at a discounted price and it sold after a month or so. The new buyer hadn't wanted to pay the $4,300 premium, either, so he had found an alternative plan from the private insurer Lloyd's of London, a British company that sells boutique flood policies to people dissatisfied with National Flood Insurance Program plans. There was no telling how reliable the policy would be, or how long Lloyd's would offer it to the new owners if the company started to rack up losses. As in Santa Rosa, there was always the possibility that a private insurer could pull out if things got expensive. But none of that was the Lanes' problem— they moved to Fernandina Beach, Florida, in July and never looked back.

The raise-or-raze conundrum almost derailed the Lanes' lives. Vernon, on the other hand, stands to benefit from the teetering flood insurance market. When I arrange to meet him for a chat, he drives me out to a neighborhood of Virginia Beach called Lynnhaven Colony. The neighborhood is a seventies-era subdivision carved out of a swampy tidal basin, each street bordered by canals and protected by a flimsy stone bulkhead. The signs of flooding are hard to miss: there are green mold marks around some of the facades, flood vents on almost every home, and patches of grass that have been blanched by salt water. The flood insurance premiums here are just as expensive as the Lanes', Vernon tells me, but the people who live here can afford to pay them.

It's here that Vernon plans to launch the next phase of his business, one that takes the raise-or-raze mandate to its logical conclusion. In addition to selling flood vents and basement fills, Vernon is now selling his clients a home elevation package. He figures out what kind of work is necessary to raise the home above flood level, farms out the hard labor to his contractors, and takes a fee for consulting. His ultimate idea is to work with a local bank on the creation of a special type of loan that will roll together a home elevation and a thirty-year mortgage, allowing new buyers to finance the cost of raising their homes up front.

Vernon drives me down a cul-de-sac where his company is already supervising five or six home elevations, including one that's in progress as we drive by. It's not a pretty sight. The construction team has already jacked the house up on its new concrete base, but they've had to gut all the plumbing and wiring in the original rooms.

I stop to talk to the owner, Tim, who is helping the construction crew with some of the work. He's drenched in sweat and his face is red—the new windows and flood vents have been installed in the wrong locations, he tells me, which will set the work back a couple of months. The elevation process has also involved ripping up a back deck and pool area, leaving the backyard barren except for a pit of rubble and a pile of wooden slabs.

"It's a bitch, that's for sure," Tim tells me, "but I don't have any other options." By the time it's finished the elevation process will cost him $400,000, around as much as his home is worth. The cost of the elevation can be seen as a hedge against an even further drop in value, the drop that would come if he tried to sell the home without tamping down the stratospheric flood insurance premium. That's an enormous price to pay to protect an investment, but Tim seems to be able to afford it: from inside the hollow of the construction site I can see a small fleet of boats that together are worth almost as much as his house.

As we drive off, I'm left wondering about the Norfolk residents who can't dump hundreds of thousands of dollars into saving their homes. The elevation gambit might help if you live in Lynnhaven Colony or on the ritzy Larchmont waterfront, but what happens to the people who can't afford to raise their homes? People like the Lanes, who live in the most vulnerable areas but can't mitigate their risk, are only going to pay more for flood insurance as time goes on, and their homes are only going to become less valuable as their premiums rise.

Who's going to help them?

III.

Saving Filer Street

K aren Speights's neighborhood is one big construction zone. The roads are stripped and lined with barricades and mesh fences. Huge pipes and slabs of wood lie on the sidewalk. There are diggers and backhoes on every block, crawling around each morning to scrape up some new street corner or deposit some new chunk of soil. Karen can hear humming and whirring all day as a crew works to build a pump station near the end of the block. Things are even louder downstream, where there are at least a dozen cranes and dump trucks gathering raw material for an enormous sandy berm that will separate the river from the neighborhood. The berm winds its way along the neighborhood waterfront and curls through the marsh behind Karen's property, bordering her backyard. It will look nice when it's finished, but for now there's a lot of noise.

Karen lives in Chesterfield Heights, a pocket community on the Elizabeth River in central Norfolk. It's a quiet place, made up for the most part of old houses like Karen's yellow bungalow. The houses in the neighborhood sit far back from the street and almost all of them have broad wooden porches, where locals tend to sit for hours at a time, calling across the street at one another or waving hello to the people who drive by. No one lives more than a few blocks away from the roar of the interstate, but no one lives more than a few blocks away from the placid banks of the river, either.

This construction project is perhaps the largest climate adaptation initiative in the United States. It is being funded by a $115 million grant from the same program that paid for the relocation of Isle de Jean Charles, the eroding Indigenous community outside the levees of southern Louisiana. Over the course of five years the city has elevated and fortified the area's low-lying streets, retrofitted its substandard drainage system, and built earthen berms along the waterfront to deter storm surge. A pair of new pump stations will push water out during heavy rains and a new tide gate will help keep it from pouring in during high tide. The crown jewel of the project is a large new park at the eastern edge of the neighborhood, one that offers residents not just playgrounds and sports fields but also a resuscitated tidal creek that will soak up even more water. The park is called Resilience Park.

If you start listening to academics and policy makers talk about climate adaptation, it won't be long before you hear someone mention that word—"resilience." This term has become more and more popular as communities stare down more frequent and more damaging weather events, but even its proponents have a hard time offering a simple definition. In physics, the concept of resilience refers to a substance's ability to spring back into its original shape after a physical shock; in pop psychology, it refers to the inner strength acquired by people who have endured adversity. In the world of climate change, a "resilient" project is anything that helps absorb the impacts of future disasters, whether those impacts are physical, financial, or social.

Just as there are many different definitions of resilience, there are also many different ideas about what the goal of climate resilience should be.

The most obvious benefit of a project like the one in Chesterfield Heights is that it will protect residents from the physical danger of severe floods; the second-most obvious goal is that it will protect them from future repair costs. The city, state, and federal government will also see a financial benefit. According to one oft-cited statistic, one dollar of resilience spending can save at least six dollars of recovery spending after a disaster.

In a deeper sense, though, the $115 million price tag for the so-called Ohio Creek Watershed Project is the cost of buying time. It is a sacrificial offering to the housing market, an attempt to beat back for a few decades the economic forces that almost drove the Lanes into bankruptcy. In guaranteeing that the neighborhood's homes remain safe from the water, the project will also guarantee that the homes remain safe from the raise-or-raze reckoning that is already approaching in other parts of the city. In a literal and metaphorical sense, it is a bailout.

From her front porch, Karen Speights has watched the resilience project alter the course of history in Chesterfield Heights. She grew up in the yellow house on Filer Street, a block that she describes as "a neighborhood within a neighborhood," but left for a few decades to attend college and start a career, moving back to the neighborhood in 2010 when her mother, Lillian, started to need help around the house. As Karen settled back into her childhood bedroom and caught up with her old acquaintances from the block, she learned that the flooding problem on the street and in the neighborhood was far worse than she remembered. The creek behind the Speights house had swallowed parts of the old family garden while she was gone, and water now seeped beneath all the homes on her side of the street a few times per month. Then a nor'easter blew through the city in 2009 and flooded the first floor for the first time in the more than eighty years that Karen's mother had lived there. Karen and her mother were sitting down to a birthday dinner of crab legs when the water started pooling around their feet.

With these devastating floods came a devastating bill. The premiums for the home's FEMA-backed insurance policy had jumped to more than

$5,000 by the time Karen moved back, and they continued to rise from there. The home was too old to elevate and retrofitting it as the Lanes had done would cost at least twice as much as the property was worth. For a while Karen thought about selling the home and moving with her mother somewhere else, but the yellow house was the only home Lillian had ever known. What's more, the house would fetch $75,000 at most if Karen put it on the market, not enough to cover the cost of moving herself and Lillian into a new place. Karen made a decent salary working for the gas utility, but she was going to retire soon, and she worried her pension might not be sufficient to keep up with another mortgage.

Even as Karen pondered whether to leave Chesterfield Heights, many of her former neighbors on Filer Street had already departed. While she was gone, the families she had grown up with had slipped away from the neighborhood one by one. The Jones family, for instance, had always lived next door to Karen, right at the edge of the river, but when the matriarch of the family died, the home fell into disrepair. Rather than pay to repair the damage from repeated flooding, Ms. Jones's children decided to move on. They couldn't find a buyer for the ruined house, so they sold it to a developer, who knocked it down and replaced it with a large white duplex elevated off the ground. On the other side of Karen's house, an elderly woman named Truler died a few years later, leaving her children to figure out what to do with the family's longtime home. The children made the same choice and sold the house to a developer for cash. Karen's across-the-street neighbor Mike, meanwhile, had seen more than half his property disappear into the creek over the decades, and had struggled over the same period to pay a flood insurance premium that was even higher than Karen's.

Had the federal government not intervened with the Ohio Creek project, Karen and the other remaining residents of Filer Street might have left as well, along with many other waterlogged homeowners in Chesterfield Heights. As the flood resilience took shape, however, it gave the neighborhood a new lease on life. The berms snaked along the shoreline, the pump stations started humming along the creek, and the flooding in Karen's back-yard began to abate. Water no longer gushed out of storm drains in reverse and the stench of sewage no longer appeared after a bad rain. The residents of Filer Street no longer had to move their cars and trash cans before the autumn high tides. The creek stopped eating away at Mike Godfrey's yard.

Lillian never got to see the project come to fruition—she passed away in September of 2022, just after her ninety-fourth birthday. Even in her moth-er's absence, though, Karen felt safe in the new Chesterfield Heights. She would be able to live out her days in the house where her mother raised her.

"It definitely had an influence on me staying," Karen told me, "because with me ready to retire in the next couple of years, I could not sustain that amount of flood insurance, and I don't know how much longer I can stay here. Now, with the project, it's indefinite—through my lifetime."

The impact of the project on the future of Chesterfield Heights is even more significant given that the project almost didn't happen. The residents of the neighborhood owe their salvation to a strange and unlikely sequence of events, a rare moment of alignment between various branches of govern-ment. If the course of history had been even a little different, Filer Street and all the blocks around it might have been consigned to the waves.

The idea for a flood protection project in Chesterfield Heights began in the early 2000s when a group of architecture students at a local university conducted a neighborhood survey about infrastructure problems. At the time, neither the students nor the long-neglected residents were under any illusion that the flooding would ever get fixed, but it was interesting to get a sense of the problems. The students drew up plans for berms, pump stations, and tidal parks, and then shoved those plans in a metaphorical drawer. Then, three years after 2012's Hurricane Sandy, as coastal cities began to wake up to the necessity of adapting to storm surge, the city of Norfolk hosted an event with the Dutch government to explore solutions to the city's flood problems. City officials, architects, community leaders, and dike experts from the Netherlands collaborated on adaptation plans for around a dozen Norfolk neighborhoods, and the participants used the architecture students' mothballed survey as a basis for the Chesterfield Heights proposal. The project was appealing, but no one thought it would ever get built. Chesterfield Heights was small, secluded, and far from an outsize contributor to the city's tax base, which meant it would be almost impossible for the city to find the money for the project.

But the federal government had also started to make tentative plans for climate adaptation in the wake of Sandy, and soon those plans bore fruit. The year before the Dutch conference, the Obama administration announced the National Disaster Resilience Competition, a grant program for cities and states that wanted to get a head start on adapting to climate change—a sequel to the FEMA program that had funded buyouts in Lincoln City and Houston. The city of Norfolk and its neighbors, hungry for infrastructure money, applied with a whole suite of potential projects, the grant equivalent of throwing pasta at a wall to see if it sticks.

Even then, Chesterfield Heights was not at the top of the list. Norfolk's most ambitious proposal was the demolition and reconstruction of the so-called St. Paul's Quadrant, a massive conglomeration of public housing projects that was also vulnerable to flooding. The projects in St. Paul's were old and dilapidated, and suffered from many of the same climate issues that Chesterfield Heights did: the buildings sat on reclaimed creek land, so the entire community flooded after even a light rainfall, with puddles the size of football fields consuming whole roadways and blocking children from walking to school. Rather than fortify the neighborhood against flooding, the city wanted to knock it down and replace it with a new mixed-income development, as many other cities had done to their public housing projects—the only difference being that here the city could justify the redevelopment as climate adaptation.

In a twist, the feds passed on the St. Paul's redevelopment proposal. It was too expensive, perhaps, or just too controversial. Instead they looked farther down Norfolk's list and zeroed in on Chesterfield Heights as a neighborhood where a resilience project could have a big impact on a low-income community for which money might otherwise be hard to find. The Chesterfield Heights project seemed to fall in a Goldilocks zone between ambitious and achievable, and with the stroke of a pen the federal grant makers saved the neighborhood from ruin. The money flowed in, the backhoes trundled down to Filer Street, and the future of the neighborhood was revised overnight.

It's easy to imagine what would have happened if the grant makers had sprung for a different project. Karen Speights would have had to move out of the leaky house, her neighbor Mike would have gone broke paying for flood insurance, and her other neighbors would have moved somewhere else after growing wary of future flooding or lonely on the abandoned street. The homes they left behind would have rotted away or fallen into the hands of developers who would have knocked them down. Elsewhere in the neighborhood, other residents would have continued to struggle with flood insurance costs, the drainage system would have collapsed, and home values would have fallen even further than they already had. The residents who didn't have the money or the strength to move out would have been forced to stick around and pray for mercy from the water.

The Chesterfield Heights project is a testament to the power of infrastructure to preserve the status quo, but it also reflects the limitations of "resilience" as a strategy for protecting against climate change. Flood walls and tidal parks are an effective way to forestall the market chaos caused by rising seas and high flood insurance, but these projects are also unjust by their very nature. If there is not enough money to protect every place in every city, then some places will always get left unprotected, and there will always be something arbitrary about the decision to fund one project over another.

Even in a place like Chesterfield Heights, resilience serves to delay rather than resolve the question of retreat. The bioswales and berms might protect Filer Street from flooding in the short term, but they will not stop the sea level from rising. What happens thirty years from now, when the water has come up another eighteen inches and when extreme rain events have become far more common? Will the federal government find the money to save Filer Street again? And if the answer is no, how long will it be before the fuse once again starts to burn down?

IV.

When the Music Stops

C onsider that there are at least eight thousand homes today in the Norfolk region that will see significant flooding over the lifetime of their mortgages. Depending on how fast the ocean rises, that number could double or triple over the next fifty years and increase tenfold by the end of the century. In other words, this midsize military town contains billions of dollars of real estate value that are all but guaranteed to vanish sometime in the next few decades. Now consider that more than half the people in the United States live within an hour's drive of the ocean. The combined value of all the coastal real estate in the United States that faces flood risk from sea-level rise is somewhere between half a trillion and two trillion dollars. That represents between 2 and 4 percent of the national economy, about the same as an industry like construction.

If the entire American construction industry were at risk of collapse in the next fifty years, you probably would have heard about it, but real estate is different. The value of a home depends not only on the materials used to build it and the neighborhood that surrounds it but also on who wants to live in it and what they are willing to pay. The housing market is not just a market for brick structures on grass lawns. It's also a place where people buy and sell ideas about happiness and visions of the future.

Sea-level rise will threaten the physical structure of homes like Karen's, but it poses an even greater threat is to the intangible value that haloes every coastal home, the magic number that buoys each property within the housing market. Just as a two-story colonial will cost more in a ritzy suburb than in the frozen reaches of Alaska, so, too, will a home in a dry neighborhood fetch more than a home in a wet neighborhood.

So it's not enough for a city like Norfolk to prevent damage to individual homes. It must keep the area around those homes livable as well, which is easier said than done when climate change is involved. As the city government discovered when it began to plan for adaptation after Sandy, there are at least a dozen neighborhoods in Norfolk that need an investment on the scale of the Chesterfield Heights project. The city and state have nowhere near enough money to fund even one of those projects, and the federal government only disburses money in haphazard and uncoordinated ways. We know how to buy time, but time is very expensive.

With far more need than they had money, city officials had to prioritize. In 2016, just as the Chesterfield Heights project got going, Norfolk released the Vision 2100 plan, a first-of-its-kind document that lays out a road map for adapting to a century of climate change. The plan includes a lot of optimistic language about "designing the coastal community of the future," and offers suggestions for how the city can transform streets and shorelines, but its main goal is to establish which communities the city will protect.

Vision 2100 divides the city's neighborhoods into color-coded hierarchies based on how vulnerable they are to flooding and how many commercial and cultural assets they have. In green zones, which sit on high ground, the city plans to build more housing and seed new investment. In red zones like the naval base, which are vulnerable to flooding but essential to the city's operations, the city will spend whatever it takes to keep things dry. The hard part is figuring out what to do with the yellow zones, areas that have high flood risk but few essential economic assets. These are residential areas where in the coming decades flooding will become an existential threat. The plan calls for structural interventions like berms and seawalls and suggests giving grants to help residents raise homes and convert lawns into wetlands, but it also notes that "some portions . . . will remain vulnerable to flooding no matter the steps taken by the City and others." Those "portions" include not only homes but also streets, water mains, gas pipes, bulkheads, telephone poles, generators, and septic systems, and countless other structural doodads.

The scale of this challenge is apparent in the largest yellow neighborhood of all—Larchmont, the idyllic area where the Langfords felt the tremors of the coming crash. Almost all the homes on the neighborhood's waterfront streets have now been elevated; the first ones were raised around eight feet off the ground, but the later elevations went up ten or twelve feet, an attempt to comply with the city's revised zoning code for new structures. Some of the elevation structures are tidy brick, while others are naked, ugly concrete; the children of one family have drawn chalk daisies all over their new cement foundation, an attempt to cover up an obvious eyesore. There are tidal puddles on all the riverside roads, even when it hasn't rained in days; saltwater fish flop around in some of these puddles, jittering over the asphalt after sloshing up during high tide. Elsewhere, bulkheads have eroded away and collapsed, and there are places where the water has eaten away the dirt that anchors telephone poles in place. On every street is "alligator pavement," asphalt that has cracked in a scale pattern from too much exposure to water; on some of the lawns, the outline of high tide has created patches of yellowed grass.

As with the flooded roads in the Florida Keys, the city of Norfolk is on the hook for maintaining all this infrastructure. Property taxes from city residents can cover the cost of routine repairs to potholes and asphalt cracks, but now the demands of the water are swelling otherwise ordinary line items in the municipal budget. It will cost a million dollars to repair the bulkhead, another few million to raise the streets, a few million more to anchor the telephone poles; the city can afford that, but the new infrastructure will only last a few years before it needs to be replaced again, at greater cost. The city could try raising property taxes on the people who live along this street, forcing residents to pay to maintain their own area, but that will only net them a few million, and it might make locals revolt at unfair treatment. Even with every house lifted off the ground, the math doesn't work. The cost of defending Larchmont against the water is more than anyone can afford.

———

Think back to the dynamite. Imagine that thousands of people in Larchmont are each holding one stick. They want to keep the dynamite in their hand for as long as they can before they throw it. Each person is watching the fuse dwindle, and each person must decide when to let go. What might make a person decide that it's time to throw the stick? Some people will wait until there's an inch or two left of the fuse, others until the fuse gets right down to the nib. Others might hold out until they see someone else let go, or until they see a few people let go at once. People are unpredictable, and the result of that unpredictability is that we have no way of knowing for sure when the property market in a place like Larchmont will start to unravel. There's no way to know at what moment the fuse on the dynamite will run out of wick.

We do, however, have some idea of what will happen once the blast goes off.

Imagine that it is the year 2038, fifteen years from this book's publication date. The sea level in Norfolk has risen by around five inches, so that the lowest-lying neighborhoods along the Elizabeth River are now just a foot or two above sea level. The city has still not seen a direct hit from a hurricane or tropical storm, and the federal government is hard at work on flood protection projects, but there are now dozens of flood events in low-lying neighborhoods every year.

The wet season this year is even worse than usual. One day, a large rainstorm drops several inches of water over the region, jamming water back through the city storm drain system and overflowing several dozen creeks. The water spills over into the streets and submerges thousands of front yards. Two days later, another storm cruises up through the Atlantic, this one arriving just as the tide is highest in the Chesapeake Bay. The wind also happens to be blowing out of the northeast, pushing even more water back into the bay to mix with the water from the first storm. The city is swamped. Water surges through the doors of retail shops and outlet malls, submerges parking lots and highway on-ramps, sits silent and frog-like over the lowest-lying neighborhoods.

The water drains out after two days, but something has changed. The homeowners in Larchmont decide they have had enough. More than two dozen of them put their homes on the market in the months after the storm.

The sellers have listed their homes at a discount—they're not stupid—but as the months wear on, they find that none of them are getting any bites. Buyers looking in the Norfolk area have seen Larchmont on the news when it floods, and not even the most convincing real estate agents can assuage their fears. Plus, the flood insurance premiums for homes in this neighborhood have risen into the thousands of dollars. The previous residents were paying a discounted rate for as long as they lived there, but new owners will have to pay the full price. The high insurance costs mean that potential buyers can't get approved for a loan unless they have an enormous amount of cash on hand, and everyone with extra cash is now looking at homes in safer neighborhoods.

As their listings expire, the flighty homeowners adopt different fallback plans. The wealthiest of them eat the loss and move somewhere else, as do those who have already paid off their mortgage in full, while others start renting out their homes in the dry season, leaving them empty during the flood-filled summer and autumn. Some of them choose to list their homes again with slashed prices, and a few of them sell, driving down the median price for homes in the area. Now other residents, ones who didn't want to move, are realizing their homes are underwater, and that they're destined to lose money on their investments. They start listing as well, hoping to get out before they lose any more.

Then the other shoe drops. A buyer makes an offer on a home in Larchmont and goes to the bank to get a mortgage loan. The bank tells him he needs flood insurance, but the home he wants to buy has flooded several times in recent years, which means the FEMA-issued flood insurance is around $12,000, so high that the buyer no longer qualifies for a mortgage. The buyer tries getting private flood insurance, but the private insurers won't touch the home. The bank's loan officer sees no reason to take a risk on a property that will likely continue to drop in value, so she tells the buyer to take a hike, and the man does take a hike—to the offices of the local newspaper. The story runs the next day, and while the details are a little fuzzy, everyone gets the general idea: a local bank has refused to make a loan in a flood-prone area of Norfolk.

In other neighborhoods across the city, neighborhoods that face the same risks as Larchmont, homeowners start to get jittery. Some in Ingleside and Ocean View decide to put their homes on the market, but they, too, encounter resistance from potential buyers. They slash their asking prices, get out however they can, and the panic grows larger. The navy announces that it may consider shifting some operations away from Norfolk, moving personnel toward areas that aren't as risky.

The ripple effect now spreads beyond the individual homeowners and begins to disturb the labyrinthine financial system that makes their mortgages possible. Large banks like Wells Fargo and Bank of America have written thousands of mortgages in the Norfolk area, but they don't seem too concerned even as home values plummet in Larchmont and some mortgage holders go belly-up. That's because the banks have already taken steps to shield themselves from a climate-driven crash like this one. Many financial institutions have already begun to conduct internal "stress tests" of their mortgage holdings, using proprietary data to identify the areas with the greatest flood risk. In many cases, when banks write home loans in one of these areas, they quickly sell the mortgage to either Fannie Mae or Freddie Mac, the two quasi-government entities that prop up the mortgage market by purchasing millions of loans per year. The result is that the federal government is left holding the bag for several billion dollars' worth of toxic mortgages.

After years of inaction, the federal government puts its foot down. Fannie and Freddie already require the mortgages they buy to have flood insurance, but now they say they won't buy any new mortgages in flood zones, period. The largest private lenders soon follow suit, refusing to assume the full risk of writing loans in areas like Larchmont. Meanwhile, major credit rating agencies downgrade the bond ratings for several small towns and cities along the Chesapeake coast. The local officials who run these towns and cities do not need to be told why. The reason is obvious. People are leaving as fast as they can.

Except leaving is easier said than done. The wealthy can cut their losses and find new homes on higher ground, leaving their condos and beach houses empty while they wait for the government to sort out the mess. People who rent their apartments are also free to leave whenever their leases are up. The ones who are stuck are the middle-class homeowners, the people who emptied their savings to buy a home and can't afford to take out a second mortgage until they off-load the first one. Some of these homeowners knew they were buying in risky areas, and many of them had flood insurance, but insurance can only go so far to protect these people from the vicissitudes of the market. Many other homeowners don't even live in the flood zone but have seen their homes plummet in value anyway as investors and employers flee the vulnerable neighborhoods around them. In literal terms, what is happening to these people is not a disaster, since it involves neither natural forces nor physical destruction. All the same, they are stuck, and there is only one entity that can help them: the federal government.

Since it just so happens to be a midterm election year, the president promises that help is on the way. Congress convenes an emergency hearing to figure out what sort of relief it can offer to the thousands of homeowners along the coastline who are underwater on their mortgages, figuratively if not literally. Some lawmakers, speaking for their helpless constituents, argue that the government has an obligation to bail these homeowners out. Others, arguing for the interests of the American taxpayer, say these homeowners made their own choices and that they cannot expect Congress to offer them a do-over.

Indeed, there's only one thing everyone agrees on. It is time to retreat.

—

A. R. Siders was at Columbia University when Hurricane Sandy struck New York and New Jersey. The storm surge gobbled up sections of the Atlantic coast overnight and flattened whole neighborhoods in Brooklyn, Queens, and Staten Island. Floodwater sluiced through the subways and the power went out in lower Manhattan for a period of days. Even all the way uptown on the Columbia campus, Siders could feel the city in a state of shock. She had come to Columbia for a postdoctoral fellowship in environmental law, but the recovery from Sandy pushed her research in another direction.

"Everyone was talking about *how* we should rebuild," she said, "and I kept thinking, *should* we rebuild?"

Siders viewed the storm as clear proof that New York and New Jersey needed to move back from the shore; it seemed obvious to her that people should not have been living in the places where the storm surge had been most devastating. Nobody else seemed to share her opinion, and she watched as the local governments in the region rushed to build everything back just as it had been. There were a few FEMA-sponsored buyouts of neighborhoods on Staten Island, but other than that, nothing.

"We were talking with some local officials, and they kept saying, oh we can't retreat, that's not something to do with the United States," she recalled. "I knew that was wrong."

Siders has since become the foremost scholar of the concept known as "managed retreat," or coordinated withdrawal from areas vulnerable to climate change. She is the author of what she calls "the managed retreat handbook," a document that catalogs several instances of whole-community climate migration in the United States. At the time that Siders created the handbook, the list of case studies she could pick from was very short. There were at most a dozen communities that had ever executed anything like a successful managed retreat, all of them small and isolated. The earliest examples include Soldiers Grove, Wisconsin, whose six hundred residents moved back from the Kickapoo River in 1978, and the town of Valmeyer, Illinois, which did something similar about a decade later. In recent years, as shoreline erosion has accelerated, the federal government has sponsored the relocation of Native American villages in Washington and Alaska, as well as of the Isle de Jean Charles community in coastal Louisiana. None of the communities that Siders wrote about numbered more than a few hundred people, and furthermore, almost none of them achieved complete relocation—some residents always chose to stay behind if they could, or else pull up stakes and move elsewhere altogether. There have been more managed retreat projects proposed across the country since then, but not many.

Seeing what Sandy did to the New York shoreline convinced Siders that larger coastal cities need to follow the example of Soldiers Grove and Valmeyer. Millions of people on the Eastern Seaboard are struggling to adapt to the sea-level rise that has already occurred in recent decades, to say nothing of the multiple feet that scientists expect before the end of the century. Erosion in Cape Cod has already wiped out beaches and begun to eat away at roadways; on the opposite coast, sections of cliffside highway in California have split off and fallen into the sea; entire islands in the Chesapeake Bay have already vanished beneath the water, rendering the fishermen who lived on them for generations homeless. The constant effects of tidal flooding are already visible not just in Norfolk but in larger cities like Boston, New York, Charleston, and Miami, where an estimated 10 percent of land area will vanish over the coming decades. No one who lives near an ocean can count themselves entirely safe.

and more vulnerable these places are, though, the more diffi-
_.. will be for them to engage in the coordinated withdrawal that Siders
catalogs in her managed retreat handbook. Even if there were a way to
convince tens of thousands of people in a city like Norfolk to agree to move
somewhere else, and even if there were enough available land to build new
homes for them to occupy, the cost of doing so would be almost unfathomable, rising into the billions of dollars. Managed retreat might be feasible
and affordable for small, isolated communities like Isle de Jean Charles, in
some cases precisely because those communities have so few resources, but
for the large cities that line the Atlantic coast this kind of coordinated effort
is a pipe dream.

But just because there won't be a *managed* retreat doesn't mean there
won't be a retreat. Up until now, the housing markets in cities like Norfolk
have largely ignored the warning signs of sea-level rise, but the coming
plunge in coastal property values will drive far more people away from the
water than any individual storm. Unlike a king tide or a nor'easter, this market correction will inflict economic harm even on people whose homes have
not suffered any damage, sharply reducing the value of the most valuable
assets they own. As a city's housing values decline, so, too, does its tax base,
which means less money to pay for basic services like trash collection and
street repair. There might be federal money available for flood protection or
climate resilience, but not enough to keep a whole city running. Among the
portion of the population that stays, there will be many who want to leave
but can't, who are tethered by their mortgages to homes they can't sell. The
only solution to this spiral of inequality will be not a relocation effort but a
bailout of unprecedented scale, one that would enable anyone who wanted
to leave to do so and would also make the city livable for those who wanted
to stay.

It's unclear whether any government, even one as rich as that of the
United States, could afford such bailouts everywhere they will be necessary.
Home buyout programs have been feasible in small towns like Kinston or

in cities like Houston where housing is cheap and federal funding is plentiful. The math is more difficult in a place like Norfolk, where home values are higher and where the risk from coastal flooding will be spread across thousands of homeowners rather than concentrated among a few. Finding a few million dollars to buy out everyone in Lincoln City was easy. Finding more than a billion dollars to buy out Larchmont will be much harder, to say nothing of the other neighborhoods where home values will drop even in the absence of actual flooding.

A few savvy politicians have already begun to draft policies that would help get people out of homes. A state lawmaker from San Diego has suggested that the government buy flood-prone homes and rent them back to their owners until a flood comes, at which point the owners could cut their losses and leave. In Surfside, Florida, a barrier-island town right next to Miami Beach, the former mayor established a "relocation fund" that would offer stipends to future residents who want to move to higher ground. Some academics have proposed government-brokered land swaps between homeowners in the floodplain and landowners in states like Nebraska or Iowa who have excess land they want to develop. The Norfolk area nonprofit Wetlands Watch has put together a proposal that would leverage private money to bail out homeowners: the city will only allow real estate developers to build on high ground if they first purchase a few homes from mortgage holders in areas like Larchmont, allowing those homeowners to make a clean escape. The developers will then flip the homes into a land trust, and when the occupants move out, the city will clear the land.

All these proposals are more cost-effective than FEMA-funded buyouts, and could help rescue homeowners with stranded assets. The big uncertainty, says Siders, is whether there will be political will to enact them on a large scale.

"It raises this question of, who do we put the onus on?" asked Siders. "Is the burden on the government to fix it afterwards, or the homeowner, or the developer? There's all these places we could put the burden, and it's not totally clear where it should be." Right now the government provides significant financial relief to homeowners who lose their homes in hurricanes and wildfires, but our system isn't set up to provide relief to those who lose their homes to a slow-moving crisis like sea-level rise, or to the market fluctuations that adjust prices to account for future climate impacts. It might seem reasonable today for the government to help those who didn't know the risks they were incurring, but what about in thirty years, or fifty years? How long should the government subsidize the risky behavior of individuals?

If we do believe that the government has an obligation to help individual homeowners in coastal areas, that belief raises deeper questions about what form of compensation it should provide. It won't just be individual properties that are lost to sea-level rise, after all, but entire communities with unique social bonds and cultural characteristics. If relocating individuals is hard, relocating whole communities will be almost impossible: outside of federal Indian law, there is no legal doctrine that enshrines a right to community or to social cohesion. Even if we do find enough money for everyone to retreat from the coastline, there is no way to quantify all the intangible value that will be lost in the process of withdrawal. The new life of a climate migrant will never be the same as the old life, and for a great many migrants it will be worse. What debt do we owe these people, and can we pay it?

Difficult as all these questions are, they all point to a far more difficult question, one that will determine not just the future of coastal cities like Norfolk but the future of the country itself.

Where are all these people going to go?

CHAPTER EIGHT

Where Will We Go?

A NEW AMERICAN GEOGRAPHY

I.

The Stakes

By the year 2100, it will have been more than a century since scientist James Hansen first testified before Congress about what experts then called "global warming," and seventy years since the United Nations' deadline for decisive climate action. No one knows what the world will look like by then, not politicians or economists or climate scientists. The range of possible outcomes on the question of climate change is so large as to be almost unfathomable.

The outcome of the next century hinges on how fast we can stop burning fossil fuels and transition to cleaner and renewable forms of energy. This is a Herculian task, one that will require public and private entities to collaborate on a rapid and fundamental revamp of all the world's industry and infrastructure. The leaders of this collaborative effort must be the world's wealthiest and most powerful countries, most of which are also the largest historical emitters of greenhouse gases.

There are some reasons for optimism. In the past few years these developed nations have taken a series of strides on climate action that would have been unthinkable even a decade ago, beginning the long and difficult work of decarbonization. The European Union, for instance, has endorsed a binding commitment to slash emissions by more than half before the year 2030, and its member states have already cut emissions by more than a quarter from 1990 levels. China, now the world's largest single emitter, says it intends to reach net-zero emissions by 2030 through a phaseout of its domestic coal industry. Even the United States, where inaction has reigned for decades, is lumbering to catch up: just before the midterms, Democrats in Congress passed a large spending bill designed to speed the transition to renewable energy, albeit one with significant concessions to the oil and gas industries. And a number of individual states from Washington to Illinois to Massachusetts have adopted their own clean power plans.

That said, talking the talk is far easier than walking the walk. Despite the ambitious commitments from high-emitting countries at the recent Glasgow climate talks, the world is still heading in the wrong direction. Coal-fired power generation reached new highs in 2021 during the pandemic rebound, and the rebound from the Russian invasion of Ukraine breathed new life into oil and gas; overall emissions rose almost 5 percent from the previous year. Even the countries that have moved first and fastest on the energy transition have fallen short of their goals—Germany, for instance, missed its most recent emissions targets by a wide margin, despite spending more than half a trillion dollars on climate action over a few decades, and the Russian invasion of Ukraine has forced many European countries to consider restarting coal plants and building new terminals to import natural gas. Meanwhile, net-zero pledges from major emitters may not stand up to scrutiny: India, for instance, has pledged to draw around 40 percent of its energy from renewables by 2030, but it has hesitated to end its deep reliance on coal power. The government of Saudi Arabia, by the same token, has pledged to reach net-zero emissions within its own borders by 2060, but still plans to export millions of barrels of oil overseas. Crucially, there is no international mechanism that can hold countries accountable for meeting their commitments or punish them if they fail to follow through.

Even as the world's governments scramble to get their act together, many of the largest private corporations are doing the same dance. Around a quarter of Fortune 500 companies have pledged to reach net-zero emissions by 2050, including giants like Amazon and Facebook. Many of the world's largest automakers plan to cease production of internal-combustion vehicles by the next decade, and demand for electric cars could eclipse demand for gas-powered vehicles well before then. The rarefied world of finance and banking is also waking up to the climate crisis, if only because it threatens the bottom lines of many bankers and investors. A number of bulge bracket banks and global investment firms have pledged that they will not finance new oil exploration in places like the Arctic National Wildlife Refuge, and large insurers are also refusing coverage to new fossil fuel projects. A growing number of hedge funds and investment firms are choosing to invest based on "environmental, social, and governance" values, on the premise that greener companies will attract more business and perform better over the long term. Given the slow pace of government action in the United States and other developed countries, it will be impossible to achieve a carbon-neutral planet without some contribution from private industry.

But these private pronouncements must be taken with an even larger fistful of salt. Several of the most ambitious corporate net-zero pledges rely on the purchase of misleading "carbon offsets," such as when a company pays to protect a patch of forest from future logging; other pledges may be little more than publicity stunts, with no real enforcement mechanisms. Furthermore, the largest companies have the most complex supply chains, and rely on countless contractors and business partners, so it may be difficult for them to reduce emissions even if their executives act. Some carbon-intensive industries like maritime shipping, meanwhile, have taken scant action on emissions, and on the banking side, there is still plenty of institutional money available to finance new oil and gas projects: large asset managers like BlackRock hold billions of dollars's worth of oil and gas shares. The financial industry may be prepared to help save the planet, but it is not yet prepared to sacrifice its own bottom line.

Achieving climate stability in the coming century will require much more than half-baked commitments like these. We will need decisive mobilization from our sclerotic governments, unprecedented altruism from our profit-driven private corporations, and active participation from a public that has many other issues on its mind. It's a very tall order. That said, equilibrium is not impossible. Right now the world is on pace for a catastrophic temperature increase of somewhere between 2.5 and 3.5 degrees Celsius, assuming every country sticks to its stated climate goals, but that increase is by no means set in stone. If we can enact strict restraints on carbon output, seed massive investment in green technology, and revolutionize emissions-intensive industries like agriculture, global temperatures could in theory plateau at about 1.5 degrees Celsius above preindustrial levels, low enough to avoid the collapse of civilization as we understand it.

If the good news is that we can hold warming to 1.5 degrees, though, the bad news is that a temperature increase of that magnitude will still bring about ecological changes that have no parallel in recorded human history. Sea levels will still rise by at least six feet, enough to inundate coastal cities and swallow oceanic islands. The unrelenting threat of extreme heat will render many parts of Central America, Asia, and Africa all but uninhabitable, scorching once-fertile areas and ratcheting desert temperatures to lethal levels. At the same time, a chain of otherworldly disaster events will ravage almost every corner of the globe—freak windstorms tearing across the American heartland, uncontrollable blazes charring much of western Australia, and a succession of monster hurricanes pummeling the Caribbean. There is no way to calculate with certainty how many people will lose their homes as a result of these catastrophes, but even a conservative estimate suggests that more than hundreds of millions of people worldwide will meet the status of climate migrants by the end of the century. In the United States alone, at least twenty million people may move as a result of climate change, more than twice as many as moved during the entire span of the Great Migration.

Climate migration is often discussed as a phenomenon that is tangential to the climate crisis itself, a secondary impact that has future implications for issues like immigration and urban planning. Nothing could be further from the truth. By the middle of the century, housing displacement will be the most visible and ubiquitous consequence of climate change, the one dynamic that unites the sinking islands of the Sundarbans with the desiccated landscapes of Guatemala and Chad. It will function as the currency of climate damage, the common consequence of climate-driven famine and climate-driven bankruptcy and climate-driven armed conflict. Climate change makes the world more unstable, and instability makes more people move.

This instability is already present in many parts of the United States, and the patterns of short-term climate displacement have already become clear in many towns and cities, but the longer-term patterns of migration are still far from certain. No one can say definitively which regions will absorb the largest share of climate migrants and which will transform to meet the demands of a remade economy. We can guess, and theorize, and make projections, but we can't be sure. The question of destination is the long shadow cast by the reality of displacement, the still-uncertain flip side of the crisis we are already confronting.

This uncertainty is even more significant because solving the climate migration crisis will require us to build stable and sustainable communities that can thrive on a changing planet. Even if we slash our emissions, overhaul our energy system, and suck carbon out of the air, we will still need to plan for climate displacement and facilitate migration to zones of lower risk, and we will still need to undo the misdeeds of the hubristic society that thought it could control the forces of nature. No matter what we do, millions of people will lose their homes over the next century, and we have a moral obligation to build new homes for these refugees, to support them as they relocate and ease their transition into their new homes.

To do that, we need to figure out where they will end up.

II.

Negentropy

One of the most important concepts in physics is the concept of entropy, or the tendency of energy to behave more chaotically over time. If you lock a bunch of bouncing particles inside a box, the patterns of their movement will only grow more inconsistent and random as they keep bouncing around. This is what the aftermath of a disaster looks like, as a bevy of public and private forces scatter a community to the wind like dandelion fluff, blowing one family down the street and another one halfway across the country.

There is an inverse concept, though, known as negentropy, or the tendency of things to become *more* ordered over time. As particles bounce around in every direction, they begin to move according to certain patterns. It is this process of negentropy that we can observe in the long-term trajectory of climate displacement.

As we have seen in the preceding chapters, the first movements after a disaster are chaotic and unpredictable. People move short distances, staying as close as possible to the places they lived before, and the sum of these movements over time creates a churn of instability in the places where disasters have struck. We don't have much data about this sort of involuntary movement, but the numbers we do have suggest that it is far more widespread than most people think. More than six million Americans have experienced internal displacement in the past decade alone, and in the summer of 2021 around one in three Americans experienced a weather disaster. There are millions of people in the United States with stories like that of Henry Arriaga, who careened around the Santa Rosa rental market after the Coffey Park fires, or Cindy Mazzola, who took a buyout after Tropical Storm Allison flooded her neighborhood in Houston, or Chuckie Verdin of Pointe-au-Chien, who moved inland after erosion disfigured his hometown. Even as disasters become more severe and displacement becomes more common, most climate-driven moves will still be local moves like these.

Over time, though, this churn of displacement will undergo a process of negentropy, cohering into identifiable patterns of long-distance migration. Rather than just finding the closest or most affordable home they can get, climate victims will begin to move to places where they believe risk is lower, or where economic opportunities are greater, or where they have preexisting social ties. They will not just be moving away *from* certain places to avoid climate danger but will also be moving *to* certain places for the benefits they provide.

One of the earliest and most vivid examples of this trend emerged after the 2018 Camp Fire, the most destructive and deadly wildfire in the modern history of California.

In November of 2019, a year after the fire destroyed the small mountain town of Paradise, more than a hundred people gathered at a brewery to commemorate the first anniversary of the disaster. Almost all of them had grown up in Paradise, which had been wiped off the map by the blaze. They had spent the previous twelve months staying with friends and family, renting apartments they couldn't afford, and tangling with insurance companies over their convoluted claims. No small number of them had also spent that time working through grief over loved ones who had died in the fire or treasured homes they had lost in the blaze. Nevertheless, the mood was upbeat. There were some tears, but the overall atmosphere was one of camaraderie, even contentment. It was the first time many of the victims gathered there had seen each other since the fire, and many people there were meeting each other for the first time.

The brewery where these people gathered was not in Paradise, or even in California. It was in Nampa, Idaho, a fast-growing suburb of Boise. The city had become a moving destination for dozens of Paradise refugees.

The organizer of the brewery reunion was Ruth Kmiecik, a realtor in the Boise area who grew up in Paradise. She and her husband moved out to Boise in 2006, before the most recent spate of wildfires and the trend of out-migration from California. Even at the time, Ruth and her husband were hearing that Idaho was booming with younger families and start-up businesses, and they decided to give it a try. When hundreds of her fellow Paradise residents lost their homes after the fire, she did her best to help as many of them as possible find a home in the Boise area. The brewery event was a natural extension of that work, especially because so many people arriving from Paradise didn't have any social ties in Idaho. The meetups were a way to make the Gem State feel like home.

Jess Di Stefano was one Paradise refugee who moved to the Boise area with Kmiecik's help. When I spoke with her and asked her how she'd made the decision to move there, her comments were revealing: She was fed up with the wildfires and the incompetence of Pacific Gas and Electric, she said, but she also knew that the high cost of real estate in California would give her and her husband a lot of leverage on the Boise housing market—in the end, she said, they had enough money left over to put their child in a private school. The other reason she left, and one she couldn't ignore, was that she thought "Gavin Newsom did a real number on California," and she wanted to live somewhere with different politics. She used the fire as an occasion to move somewhere where people shared her values.

Another refugee from Paradise who ended up in Boise was a former political blogger named Michael Orr, who struggled for years to overcome the trauma of the fire. Orr wrote an action novel, *Burn Scar*, that recounted the death and rebirth of a fire-stricken town much like Paradise, but after finishing the novel he decided he didn't want to stick around for the real-life recovery. He started applying for jobs all around the country, and eventually he got a hit in Boise, so he decided to visit Idaho and check the place out, driving through the Sierra Nevada and out into wide golden farmland. "The second I looked over the valley from that pass," he told me, "I knew this was home—I just had to find a house." The job fell through soon after, but Orr was undeterred—the insurance payout on his house in Paradise was more than sufficient to cover the cost of moving to much-cheaper Idaho, and soon he found a large home in a small village on the outskirts of the Boise metro area, surrounded by lakes and streams that have never been touched by fire.

The trend of migration from California to Idaho is among the best illustrations of how climate negentropy will work over the next decades. Jess Di Stefano and her husband had a push factor that caused them to leave Paradise, but they also had a set of pull factors that drew them to Idaho. The area was affordable, it was only a few hours' drive away, and its climate was similar to that of Northern California. There were still wildfires in Idaho, and smoke passed over Boise during the worst of the fire season, but it was nothing like Paradise—there was no denying that the risk of disaster was much lower. Thus the migration that followed the Camp Fire fell midway along the spectrum between voluntary and involuntary. Many of the people who left Paradise for Boise might have ended up moving even if there had never been a fire, but the disaster exerted a strong influence on where they ended up and how they got there. They had the means to choose where they ended up, and a cushion that softened their landing.

The same risk-reduction logic that drove the Paradise refugees to Boise also prevailed for the voluntary migrants who left the Florida Keys after Irma. The most important thing for these departing snowbirds was to find somewhere outside the path of hurricane risk, somewhere they would not be exposed to the unmitigated wrath of a Category 4 landfall. That logic led some couples to move to mainland Florida, where they could sleep a little more soundly, but it led others like Connie Faast to move as far away as Asheville, North Carolina, a more temperate place that sits hundreds of miles from the ocean.

For those who cannot relocate to zones of lower risk, on the other hand, post-disaster displacement can mean exposure to future disasters. This was the case for many other refugees from the Camp Fire, especially those who did not have the resources to buy new homes. The blaze plunged hundreds of people from the Paradise area into homelessness, creating a semipermanent residential encampment at a parking lot in the nearby town of Chico, a place that had already had a severe housing crisis and a burgeoning unhoused population. The fire also destroyed several mobile home parks, and these parks were never rebuilt, which cast hundreds of low-income residents to the wind. Something similar happened in Lake Charles, a petrochemical hub in southwest Louisiana, where not one but two major hurricanes struck over the summer of 2020. The city remained in shambles for much of the following year, and hundreds of homes were still derelict the next summer—approaching the city on the highway you could see hundreds of blue tarps stretched out to cover holes in the roofs of houses that might never be rebuilt. Many renters struggled to find new apartments in the area after the storm, and found themselves forced to relocate to nearby cities like New Orleans; the following year, during Hurricane Ida, many of them had to evacuate back to Lake Charles. Meanwhile, the Lake Charles families who moved to safer cities like Dallas enrolled their children in the local schools and never returned to the landscape of blue tarps.

In other cases, long-term migration trends may have less to do with escaping climate risk and more to do with finding social and demographic ties that ease the process of relocation. This was the case for the enormous out-migration from Puerto Rico in the aftermath of 2017's Hurricane Maria. Refugees from the island landed in all fifty states after the storm, picking up hotel room rentals and short-term leases wherever they could find them, and at the beginning it looked like there was no rhyme or reason to this exodus.

Over time, though, patterns began to emerge in the Maria diaspora. The biggest destinations were Orlando and Miami, which was predictable: these two cities had large Puerto Rican and Spanish-speaking populations as well as ample spare housing in the city and suburbs, and they were only an hour's flight from Puerto Rico. The next two most common destinations, though, were less intuitive. They were the cities of Buffalo, New York, and Springfield, Massachusetts, both of which were hundreds of miles away from the island and had a far smaller stock of hotels and affordable housing. But both cities boasted large Puerto Rican communities that had bloomed over generations as Puerto Ricans arrived to work manufacturing jobs, and the thousands of Maria refugees who made their way to these cities after the storm found familiar community and social support. Many of these families assumed they would go back to the island within a few weeks, but a large percentage settled down in Buffalo and Springfield for good, taking jobs at truck-driving companies or in the local casinos.

The other major factor that will influence the shape of the coming century's climate migration is the same one that has influenced migration in the United States for the past two centuries: urbanization. More than 80 percent of Americans live or work in a metropolitan area, up from around 45 percent a century ago, and the coming era of climate displacement is likely to perpetuate this trend. This is not only because cities offer more jobs and more houses but also because rural areas have a much harder time recovering from disaster events. We saw for instance how Hurricane Floyd devastated the small towns of the Neuse River watershed, and how many former residents from Lincoln City had no choice but to move closer to town or to find an apartment in larger cities like Charlotte. As local and state governments make difficult decisions about where they want to spend money on flood protection and climate resilience, the logic of the market will dictate that these governments protect the most populous places, where home values also tend to be higher. In rural areas like Pinal County, Arizona, meanwhile, climate impacts on industries like agriculture will devastate local economies, forcing many people to seek other jobs in more urbanized areas. The gap between cities and small towns will only grow wider.

These patterns match up with the national projections created by Mathew Hauer, a demographer at Florida State University who has created one of the only existing models for domestic climate migration in the United States. Drawing on reams of census data, Hauer extrapolated the future course of migration away from coastal areas, hoping to figure out where people would go as they tried to escape rising sea levels. He found that the younger and wealthier members of a given community will tend to leave first, and that these people will tend to move to the nearest large city that does not face extreme climate risk. This means that people from New Orleans might move to Dallas, people from Miami might move to Orlando, and people from Mobile might move to Atlanta. This migration trend would threaten the future growth of the coastal cities where sea levels are rising, but it would be a boon for the midsize inland cities, especially those in the Southeast like Atlanta.

Like everyone else, Hauer is making educated guesses, but there's an intuitive logic to his predictions. People tend to move short distances if they can, and they also tend to move where they think they can get a job. The midsize inland cities he highlights will be ideal destinations for migrants of all income levels; for low-income families who have lost a job or a home, they will offer robust labor markets and more plentiful rental housing, and for wealthier people who leave the coast of their own volition, these cities will offer higher-end luxury homes and a more attractive lifestyle.

For most climate migrants, though, the process of finding a new home will involve balancing all these factors and more. Putting down roots in a new place is never just about finding a job or avoiding physical danger—it's always a process of compromise between values and circumstances. That process looks different for everyone.

This idiosyncrasy is exemplified by the story of Sally Cole, who left New Orleans after Hurricane Katrina. Sally was a professor at Delgado Community College in the years before the hurricane. Her students were from some of the city's poorest neighborhoods, but she herself was a lifelong resident of the city's cushy Lakeview area. In the first month after the storm, she bounced around the country between several different couches and spare rooms: she stayed with a friend in Baton Rouge, then with her son who was working as a journalist in Washington, DC, and then in New York for a little while with some other family member.

When Sally returned to New Orleans, she found that her old home was uninhabitable, having absorbed ten feet of water when a nearby levee along Lake Pontchartrain had burst. She took up residence in a trailer, and lived there for a few months as classes resumed at Delgado, but already she had made up her mind to leave the city. She was losing hours of sleep every night worrying about when the next storm would come, and besides, New Orleans felt different, sicker somehow. Her neighbors were all leaving, and many of her former students had never returned to the city after the storm—like thousands of other refugees, they had ended up in other cities and found themselves without the means to get back home.

Sally later read a newspaper article in the New Orleans *Times-Picayune* that mentioned one of her students, a young man named LeMoyne Reine, whom the article said was still stuck in Houston, without the resources to return to his hometown—"[The] future looks grim for some left to fend for themselves," the subtitle said. Most people like LeMoyne never planned to stay in Houston over the long term—New Orleanians like him were being blamed for a rise in murders, denied job interviews because of their 504 area codes, and referred to as "illegal immigrants" at public meetings. Nevertheless, thousands of Katrina refugees ended up staying in the city, either because they eventually achieved stability or because they did not have a home in New Orleans to which they could return.

Sally put in her retirement notice at Delgado a few years early and decided to start fresh. Her parents had passed away some time earlier, leaving to her and her siblings a cottage house in the woods outside Flagstaff, Arizona. No one was using the house at the time, and Sally figured it was about as far away from hurricanes as she could get. She packed up her things and drove out there around the first anniversary of the storm, hunkering down at eight thousand feet above sea level. By the time spring came around she had decided she did not want to spend another winter shoveling snow out of her driveway and gathering firewood for the cottage's wood-fired stove. She packed up again and went house-hunting until she settled on a one-story brick bungalow in Tucson, a larger and warmer city where she could have a more convenient retirement.

Unlike Jess Di Stefano and her neighbors from Paradise, Sally might never have left New Orleans were it not for the storm, but like Jess, she was fortunate enough to choose where she landed. It might seem odd that someone whose life had been so upended by the weather would choose to live in one of the hottest and driest cities in the United States, but the decision has its own sympathetic logic. Sally's desire to avoid future storms was more emotional than calculated, and the tail risks of extreme heat and drought seemed far less real than those of a tropical cyclone. She doesn't mind collecting rainwater during the driest summer months, and her yard, like all her neighbors' yards, has been planted with a desert-adapted rockscape.

Sally's aim in moving out west was not to avoid all climate risk, but to reduce her exposure to one kind of risk. She was so committed to avoiding floods, in fact, that for the first few years she lived in Tucson, she purchased flood insurance. Her home was not near a flood zone and Tucson's annual rainfall is one-third the national average, so if anything, she should have worried about having too *little* water—the city was under the same groundwater restrictions as Casa Grande, and it, too, drew water from the shrinking Colorado River. She had operated under the time-tested assumption that it was better to be safe than sorry, but it was impossible to be safe from everything.

III.

Out of the Frying Pan . . .

An important disclaimer: the largest migratory trends in the United States right now have little to do with climate change. In fact, these movements are the opposite of the ones you might expect to see in a nation where climate impacts are becoming more severe. As the oceans rise, the number of people living near them is rising as well: almost one hundred million people live in coastal counties, 15 percent more than did at the turn of the century. The fastest-growing cities in the country, furthermore, are also the hottest and driest cities—population growth has exploded in places like Phoenix, Miami, and Dallas even as it has sputtered out in more temperate cities like Detroit and Minneapolis. Regional migration patterns show a similar shift from north to south: after the most recent census, southern states like Texas and Florida gained congressional seats to reflect their growing populations, while Rust Belt states like Illinois and Ohio lost seats to account for population loss. The lone exception to this trend of movement toward risky areas was California, which lost population thanks to the wildfire crisis and exorbitant housing costs.

These background trends make it difficult to tell whether the United States is yet witnessing anything like a regional climate migration. If ten people leave Phoenix because of the severe heat, but twenty more people move to Phoenix because they got a job there, it will look like Phoenix is growing. People move for a variety of reasons, and for most people who haven't suffered through a disaster, climate change is still something they keep in the back of their minds, or maybe something they don't consider at all. It may have some secondary influence on their decisions about whether to move or where to move, but it isn't the primary reason.

This ambiguity makes it difficult to know for sure how much voluntary climate migration is happening in the United States, but recent research suggests that the phenomenon may be more widespread than we think. A survey of more than two thousand people conducted by the real estate company Redfin found that almost half of those planning to move in the next year were motivated by natural disasters, while more than a third were motivated by rising sea levels. Moreover, three-quarters of all respondents said they would hesitate to buy a home in an area threatened by climate change, even if it were more affordable. Another study, from Cornell University, found that more than half of respondents in a nationwide sample said that climate change would have at least a "moderate" effect on their decision about whether to move over the next decade. These statistics tell us that while climate anxiety may not be a direct driver of movement, it does narrow people's ideas about where they want to live or where they *would* move. It may be a while before we see the large-scale effects of this shift, but the first signs are already beginning to appear.

Take the story of Oliver and Kristyna Hulland, a couple who moved to Connecticut in 2017 after a prolonged discussion over where they would be insulated from climate impacts. Oliver was a medical student seeking a new residency and could live anywhere there was a hospital, but his wife had long nursed a dream of starting a farm, so the couple had to find a place where they would be able to grow crops long term.

"This struck most of the West Coast off the table due to water scarcity," Oliver wrote me, and "also meant the southwest and Florida were off the table due to flooding and intermittent extreme weather. . . . We looked at 50-year models both for climate change and sea-level rise, as sea levels can dramatically influence the salinity of local aquifers." By the time they factored in all these potential dangers, they had narrowed the search to the Northeast and the Midwest. They turned to Connecticut, where Oliver had grown up—"I had initially written it off but my parents are here, and we were starting a family so it made more sense," he said. The state also offered subsidies to revive its dormant agriculture industry, making the finances of a new farm easier than they would have been elsewhere.

If the Hullands' story seems like an edge case, that's because it is—the leading edge. The hyper-selective criteria they used to find a present location for their farm are the same criteria you might use if you wanted to pick a future home for your children or grandchildren, or at any rate if you wanted to choose a place where they could live without fear of climate disaster or resource scarcity. The floods and fires discussed in the previous chapters will be among the main drivers of this elective migration, even for people who don't experience them directly—with every new disaster, more and more people will reconsider their options. They will ask, "Why would I move to Houston after seeing what that hurricane did to it? Who would want to live in Phoenix after seeing how they have to ration their water?"

The largest driver of voluntary migration in the coming decades, though, will be a far more familiar phenomenon: heat. Even with drastic action on emissions, temperatures almost everywhere on Earth will continue to rise over the coming century, creating profound changes in seasonal climates on every continent. These changes will be most drastic in polar regions, which are warming several times faster than the global average—permafrost melt in Canada and Alaska has already caused massive land collapses, and a heat wave in Siberia during the summer of 2021 caused wildfires on land that was once too cold to burn.

Even in temperate regions, the changes will be tangible. The moderate temperature zone that scientists call "human climate niche," which in the United States now stretches from South Dakota to the Sunbelt, will shift northward so that by 2070 its northern edge reaches into Canada and its southern edge ends around Kentucky. The areas below that niche will get hotter and more humid with every passing year, and as time goes on, they will start to seem more dangerous and less attractive. The already sweltering South will get even hotter, the temperate parts of the country will no longer feel as temperate, and the frigid reaches of the North will feel a bit more hospitable. Those changes might not feel like much from year to year, but over the decades they will begin to add up.

It is heat that will cause climate migration to expand from a set of isolated displacement events into a phenomenon that is truly regional in nature, a demographic shift that enfolds the entire country. The risk of climate disasters like floods or fires is concentrated in specific parts of specific regions. It's not the entire city of Santa Rosa that sits in the wildland-urban interface, for instance, but only the section that abuts the mountains; it's not the entire city of Norfolk that will go underwater, but only the segment closest to the Chesapeake Bay. The same cannot be said for the threat of extreme heat: as the temperature niche shifts northward, entire states and regions will have to adjust their lifestyles and activities to accommodate both extreme heat waves and constant summer sweat.

Many of the places that will be hit hardest by this overall temperature increase are the same inland cities that demographers like Hauer believe will absorb millions of climate migrants from the coast. This is largely owing to a phenomenon known as the heat island effect: materials like asphalt, concrete, and metal trap the sun's heat as it pours down during the day, and at night these same materials release that stored-up heat back into the air, robbing residents of the cool reprieve that comes after sundown in rural areas. At the same time, many urban neighborhoods lack trees and foliage that soak up humidity and provide crucial shade cover. The heat island effect means that the ambient temperature in a major city like Atlanta can be up to seven degrees hotter than in the surrounding countryside. The uptick in temperature has already made some activities intolerable in southern cities: in 2019, for instance, the Texas Rangers replaced their outdoor baseball stadium in Dallas with a $1.2 billion air-conditioned complex, for the simple reason that fans could no longer stand to watch games in the summer. Even as people from the coast and the countryside move to these cities to escape severe weather disasters, they will have to face a mounting seasonal crisis.

As with hurricanes and drought, the burden of this temperature shift will fall hardest on those with the fewest resources. Research has shown that the wealthiest neighborhoods in any given city tend to be the ones with the least exposure to extreme heat, thanks both to lush tree cover and access to public parks. Wealthier households are also much more likely to have air-conditioning, which during hot spells can mean the difference between life and death. Consider the infamous 1995 heat wave in Chicago, a city where many residents had previously been able to get by without air-conditioning and where local officials had no plan for an extreme heat event. The lack of preparedness among households and the city government led to more than seven hundred deaths over a period of five days. Most of the victims were either elderly residents who lived alone or children in low-income neighborhoods whose immune systems shut down in response to the heat stress.

The Chicago heat wave was among the deadliest in US history, but it didn't lead Americans to view the Windy City as a place characterized by extreme heat. By the same token, it won't be a single heat event that spurs the trend of northward migration, but rather the long-term development of an oppressive new normal. Recent research from the nonprofit First Street Foundation found that the coming decades will see the emergence of an "extreme heat belt" stretching up the middle of the country from Houston to Indianapolis; people who live inside this belt can expect to see at least one 125-degree day per year by 2050, and dozens of hundred-degree days. As residents of these cities find themselves enduring the same extremes every summer, they will come to see extreme heat as a feature of ordinary existence in the places they live, which will affect their moving decisions. Some sensitive populations may move earlier than others: young children's bodies are not as good at regulating temperature, which means they are more vulnerable to heat cramps and dehydration episodes that impair cognitive function. Elderly people also face much higher risk for heatstroke, not least because many of them live by themselves and don't have adequate supervision when the thermometer gets high. Elevated summer temperatures are also correlated with episodes of aggression in adults and with increased rates of violent crime.

It's not just a matter of health, either. As temperatures get higher, so will demand for air-conditioning and water, which in turn will cause the price of those utilities to rise, making bills unaffordable for many low-income families. The increased demand for air-conditioning in the summer will also threaten the stability of neighborhood power grids and lead to sudden and extended blackouts. Large-scale power outages are isolated and infrequent so far, at least in major US cities, but they are only going to become more common as the electrical grid makes a bumpy transition from fossil fuels to renewables. It's only a matter of time before many people decide they can't afford the biological and financial burden of a summer under sweltering heat. Floods and fires are profound and traumatic events, to be sure, but a rise in ambient temperatures will degrade the quality of daily life, making average existence more difficult and unpleasant with every passing year. Risk may be a strong push factor, but certainty is even stronger.

———

It won't just be individual households moving to escape climate danger. The escalating scale of global warming impacts will also jeopardize the future of many prominent industries, from agriculture to livestock to energy to tourism, forcing a concurrent migration of business and corporate investment. In all likelihood, people will move first, but it won't be long before the money starts to follow.

Among the first industries to feel the pressure will be livestock, a sector with a double exposure to high temperatures. Extreme heat can be unhealthy or even fatal for cattle, swine, and poultry; if you've ever seen footage of factory farms that house thousands of cows or chickens in a single sweltering room, you can imagine how dangerous a heat wave would be for the profits of a company like Tyson or Smithfield. Additionally, raising and slaughtering livestock is one of the most resource-intensive activities on the planet. The average cow drinks as much as twenty gallons of water per day, whereas the average human drinks only half a gallon, and keeping swine and poultry healthy demands an enormous amount of electricity for air-conditioning in the summer and heating in the winter. These resources will become more expensive as temperatures continue to rise, which will further increase the financial threat for livestock companies in the hottest areas.

The same goes for agriculture, and not just in the arid areas like Pinal County, Arizona. An enormous portion of the land area in states like Mississippi and Arkansas is devoted to farming staple crops such as corn and soybeans, making the Southeast just as important to the nation's food supply as the more temperate Midwest. Both of these crops are very heat-sensitive and can only grow to full health within a narrow temperature range: in the case of corn, for instance, temperatures above 95 degrees Fahrenheit make photosynthesis much less efficient and also decrease pollination volume, so that the crops grow slower and reproduce less. The summer temperatures in states like Alabama are already bumping up against the upper end of that ideal temperature range; as the world keeps getting hotter, the crop harvest in these states will get weaker each year, an effect compounded by the threat of droughts and devastating storms. According to one estimate, average annual yields for corn and soybeans in the South could fall by as much as 20 percent over the next decade, leading to losses of more than half a trillion dollars.

Even more disturbing is the risk that this heat poses to people who work outdoors. More than fifteen million workers in the United States have jobs that require them to spend some amount of time outside, and many of them are migrant laborers like the cherry pickers in Washington who worked through the "heat dome" that scorched the Pacific Northwest in 2021. Heat is already the deadliest climate-related killer in the United States, and it has killed at least four hundred people on the job over the last decade—likely a severe undercount, since heat exposure often causes other symptoms. As extreme temperatures become more common in southern states, mortality rates in outdoor work will also increase as more laborers pass out from sunstroke or suffer heart attacks and strokes. The major companies that do business in these states will have a harder time finding laborers willing to work in lethal heat for meager wages, and a surge in heat deaths could also lead to lawsuits or punitive action from labor regulators.

Rather than adapt to these difficult conditions, many of these companies will eventually pick up stakes and move, especially if temperature increases are also making northern areas more suitable for industries like livestock. This has already begun to happen for large conglomerates like Cargill, the nation's biggest beef producer, which in 2013 shut down its production plant in Plainview, Texas. The area had become too hot and too dry to raise cattle, and beef producers in the area were fast depleting the southern portion of the all-important Ogallala Aquifer, a massive groundwater source that runs down the center of the country. Cargill laid off thousands of people from the plant and told them they could reapply for jobs at the company's other locations in Kansas and Nebraska. The following year, Nebraska displaced Texas as the largest cattle-producing state.

The economic carnage will only expand from there. Industries like mining and construction also require employees to spend much of their time outdoors, and these industries, too, will become more dangerous as temperatures rise. Meanwhile, the threat of summer heat waves will also contribute to a decline in soft industries like tourism—a five-degree increase in average summer temperatures could mean a drop in revenue for outdoor theme parks like Walt Disney World and Universal Studios, not to mention for regions like Palm Springs or the Grand Canyon whose tourism appeal depends on sun-scorched attractions. The data centers and other tech facilities that have opened up in Arizona over recent years also require enormous amounts of water, which may discourage future investment in parched southwestern states. Meanwhile, many large oil refineries and petrochemical facilities sit along the Gulf of Mexico, where hurricanes are most frequent and most dangerous; the drumbeat of damage from these storms may encourage oil companies to move their operations elsewhere, as happened after Hurricane Ida when Phillips 66 shuttered a much-damaged refinery in southeast Louisiana. In almost any industry that doesn't center around an air-conditioned office, climate change will mean increased danger and decreased profit.

Thus the long-term result of this shift will be a gradual migration of major industries away from the rural and resource-dependent parts of the South over the course of the next century. This capital flight will reinforce existing trends of urbanization, which could lead to further economic decline and stagnation across much of the southern part of the country. In the low-population areas that rely on single industries like agriculture or meatpacking, the pain will be severe, comparable to the coal bust that pushed parts of Appalachia into a state of permanent underemployment. There will still be millions of jobs, of course, but the region will no longer be the motor of the national economy, and people outside large cities will have fewer reasons than ever to stay put where they are. Major metropolitan areas like Atlanta and Phoenix will be better able to weather the economic headwinds, but even they aren't invincible. These cities ballooned for decades thanks to a steady influx of young professionals and southbound retirees, but fifty years from now the world will look very different. If climate change makes a city like Dallas unlivable or even just unpalatable, it won't be long before white-collar businesses open new offices elsewhere, seeking to attract talented recruits who are picky about where they live. Even the most robust urban economy isn't immune to a long-term shift in perception.

None of this is to say that the hottest and most vulnerable parts of the country will be abandoned by the year 2100, or even that they cannot continue to grow for decades to come—as we have seen, modern society can go to great lengths to bend nature to its will, and we haven't yet exhausted our bag of adaptive tricks. Even so, the fact remains that by the end of the century many of the country's most populous and industrious areas will be less fit for human habitation than they are today, and that transformation will bring with it a tremendous cost to human life and economic productivity. Unlike the recessions and crises of centuries past, this downward spiral will be one from which there is no real hope of recovery, at least not on time horizons that humans tend to use.

A permanent regional crisis like this one, should it come to pass, would threaten an ideal at the heart of American history—the ideal of expansion. The history of the United States since the very moment of its colonization has been a series of attempts to chart new lands, till new sections of ground, and build in new places. The coming era of climate catastrophe will force us to backtrack this centuries-long pattern of growth, shrinking back from the most hostile and vulnerable areas. When that happens, the eyes of the nation may once again shift north.

IV.

Safe Havens?

It may be true that every place in the United States faces some degree of climate threat, but just as there are some places where that risk is acute, there are other places where that risk is comparatively low. Just as no one would say that Birmingham, Alabama, faces the same degree of hurricane risk as New Orleans, no one would say that Madison, Wisconsin, faces the same degree of heat risk as Birmingham. These statistical differences don't matter when a hurricane strikes Birmingham or a heat wave strikes Madison, but they do have an impact on where people choose to move, settle down, and raise their children.

Thus even though there is no such thing as a "safe haven" from climate change, there are some places that will emerge as consensus destinations for climate migrants, or else as sensible sites for future investment by corporations and home builders. The most climate-resilient cities will be situated in temperate environments, insulated from rising seas, and possessed of ample fresh water.

At least on paper, that sounds a lot like the Rust Belt, and in particular the cities that border the Great Lakes. Many Rust Belt cities have been losing population for decades, shrinking down to a fraction of their mid-century size, so they have plenty of room for new arrivals. Indeed, many of these cities are already trying to brand themselves as climate havens, even if this branding effort has arrived decades early. Take, for instance, the city of Buffalo, New York, renowned for its bitter winters and stunning snowfalls. The mayoral administration there has launched a campaign to brand Buffalo as a "climate refuge city," one that is "stepping up and preparing to welcome this new type of refugee," a campaign perhaps inspired by the wave of refugees that arrived after Hurricane Maria. The chief sustainability officer for the city of Cincinnati has also said he is working to make the city a "welcoming place for people fleeing disasters and extreme heat," citing new demographic models like the ones Hauer has produced. An architecture professor named Jesse Keenan has created a proposal to rebrand the city of Duluth, Minnesota, as "Climate-Proof Duluth," attracting heat-avoidant migrants to this frigid city on the shores of Lake Superior.

The transition might not always be smooth, of course: members of the Fond du Lac band of Chippewa near Duluth worry that the "climate solution [could] end up in our cultural and spiritual genocide" if people rush to the area and push out those who have lived there for generations. There is potential in any climate haven for another phenomenon that Jesse Keenan has studied: "climate gentrification," or housing displacement that results from a place being *too* appealing from a climate perspective. There's not much evidence of this happening in the north so far, but if thousands of young people decided to put down roots in Cincinnati or Buffalo, it's a safe bet that many low-income residents would soon see significant upticks in rents and property taxes.

It could take generations for cities like Buffalo and Duluth to benefit from shifting attitudes about climate change, but by the middle of the century it is very likely that at least some of these places will be seeing far more new development and investment. Think back to Mathew Hauer's climate models, which predict that younger and wealthier residents will be the first to move away from flood-prone coastal areas. Imagine that this dynamic continues to hold for the next couple of decades, but that by the middle of the century these young migrants are no longer content to live in blazing-hot inland cities like Dallas and Atlanta, and that now they want to move to more temperate places.

The net migration trends slip in the opposite direction, the economic effects compound on each other, and soon cities in the Midwest are seeing runaway growth of the same kind that places like Austin and Phoenix are seeing now.

It's hard to know for sure what the scale of this regional migration will be, or when it will take shape, but it is almost a certainty that by the end of the century there will be a substantial flow of people toward relative safe havens and away from the riskiest and least hospitable places. It may take decades for these demographic flows to coalesce, but barring a profound shift in climate trends or government policy, many people *will* move north-ward. The driving force behind their migration will be the same force that has stacked up houses on eroding coastlines and bent the water system in Arizona out of shape to accommodate new development. Our profit-driven society relies on economic growth, and the climate crisis will make some areas far more conducive to growth than others. That's not to say vulnerable places in the West and along the Gulf Coast will be abandoned in their en-tirety, but only to say that the perpetual motion machine of capitalism will need to look somewhere else, somewhere less exposed to environmental risk and better positioned to expand as the years go on. In all likelihood, that somewhere will be above the Mason-Dixon Line.

—

In November of 2020, after that year's record-breaking hurricane season had blown through the regular alphabet, a pair of Category 4 storms slammed into Central America one after the other, killing at least one thousand peo-ple across the region. The destruction was horrific. Rapid mudslides buried entire villages in Guatemala beneath amber-tinted sludge. Bulging rivers in Honduras snatched children out of their homes and swept them away to their doom. A truck carrying more than a dozen travelers plunged off the side of a mountain eroded by rain. In many villages and towns the flooding was so severe that rescue workers could not reach affected areas for days or weeks. Even months after the storm there were towns in Nicaragua that were still submerged beneath standing water or cut off from the rest of the world by broken bridges.

A constellation of aid groups rushed into Central America to offer short-term relief, but nothing could be done to address the wider devastation. Agriculture fields were damaged beyond use, and thousands of residents had seen their entire villages wiped away, leaving them with no shelter, no money to find shelter, and no jobs that could help them make money. For the victims of this compounding disaster, the United States was the only viable place of refuge. The US had seen its own share of record-breaking storms and fires over the previous year, but most of the country was safe, untouched, whereas in Nicaragua and Honduras there were few places that had escaped nature's wrath. For millions of people living in these countries, the decision between leaving and staying was the decision between life and death—in other words, not a decision at all.

With no other options, hundreds of thousands of refugees from Central America made the thousand-mile trek north to the United States, where president-elect Joe Biden was about to take office, having pledged to revise the harsh immigration policies of his predecessor. This sudden migration surge echoed the one that had occurred in 1998 after Hurricane Mitch, the deadliest hurricane in modern history, when almost a million people fled from Central America to the United States.

If the climate crisis in the United States is still in its larval phase, then the crisis in Central America is already bursting out of its chrysalis. Climate change is not the only reason that refugees are fleeing these countries for the United States, but a succession of droughts and devastating storms has pushed far more of them northward than would have moved otherwise. Although these migrants are separated from the United States by legal and political borders, the migratory movement from Central America exists on a continuum with the churn of displacement that is beginning stateside.

If the United States has struggled to help internal climate migrants, its track record with international climate migrants has been even worse. The

Biden administration has taken few steps to roll back the restrictive immigration regime that emerged during the Trump presidency, and climate change itself has been all but absent from conversations about the border. In the autumn of 2021, Biden's White House released a brief and little-noticed study that summarized the federal government's options for responding to international climate migration. The report examined the possibility of expanding programs like the Temporary Protected Status authority, which allows the president to suspend immigration restrictions for countries affected by short-term disasters, but its ultimate recommendations were rather muted: it suggested that the government "establish a standing interagency policy process on Climate Change and Migration" and that it "assess investments in predictive tools that forecast conditions correlated with migration and displacement." Given that the climate crisis may drive millions of people to seek refuge in the United States over the coming decades, these wait-and-see proposals were far from adequate to the scale of the phenomenon.

Things are little better across the Atlantic, where the European Union plays a parallel role as the closest destination of refuge for migrants from Africa and the Middle East. The influx of a million refugees to the continent in 2015 set off a chain of events that fractured the EU, leading to the withdrawal of the United Kingdom from the bloc and the rise of far-right parties in several countries. This political turmoil will only intensify as the Middle East and North Africa see prolonged drought, killer sandstorms, and daytime temperatures of 120 degrees Fahrenheit. Meanwhile, the potential for long-term climate upheaval may be even greater in Central and Southeast Asia. Already island nations like Tuvalu and Vanuatu are planning to relocate their entire populations away from rising sea levels, while transformative shifts in soil ecology are pushing fertile farmland north from China into a warming Russia. A heat wave in India during the summer of 2022 exposed a billion people to triple-digit temperatures for multiple weeks, while flooding in China during the same season displaced hundreds of thousands of people around the province of Guangdong. Most climate-related displacement is still internal, but as time goes on the overall number of international climate refugees will continue to grow. Compounding disasters in these areas will drive yet more movement toward the very countries that have already demonstrated their extreme hostility toward migration from the Global South.

But focusing on the potential scale of movement toward the developed world may be the wrong way of approaching the problem. The climate journalist Alexandra Tempus, writing in *Rolling Stone*, has argued that fearmongering about hordes of displaced migrants may encourage developed countries to close borders and restrict refugee flows, especially given that these countries themselves will be dealing with their own climate disasters—"Apocalyptic predictions may grab our attention," she writes, "but they can also stoke xenophobia and miss the full picture of what's happening on the ground."

Given that these rich countries are the same historical emitters who warmed the earth in the first place, the moral response for a nation like the United States is not to close its borders but to rethink how it provides housing, both to its own citizens and to those arriving from abroad. Domestic and international displacement are very different political challenges, but the solution to both is the same: in an age of permanent instability, we must do everything we can to create a universal guarantee of shelter. And if the impacts of climate change are global, our response to those impacts must be global too.

V.

A Question of Responsibility

H ow we adapt to the era of climate displacement will depend on how we answer a very simple question: What do we owe to each other? This question is less about the ethical obligations of any one individual to another and more about what we as a society owe to those on whom the impacts of climate change will fall the hardest. How much money should taxpayers spend to guard coastal towns from rising seas, to rebuild neighborhoods that get destroyed by fire, to bail out farmers who don't have enough water? Does the government owe its citizens the freedom to live wherever they want, no matter the risk, or does it have a responsibility to move them out of the most vulnerable places? In places like Louisiana and Arizona, places where the status quo is untenable, who has a right to remain?

We have seen that these questions are most vivid in the immediate aftermath of major disasters, when the government must work as fast as possible to make victims whole again. This task is impossible almost by definition, but there are clear ways the government can improve. The hierarchy of federal disaster policy is sluggish and outdated, focused for the most part on rebuilding properties where they once stood. The government by and large does not try to track down the thousands of people who vanish after these events, nor does it offer adequate long-term support to people who have relocated.

The first step to addressing displacement, beyond reducing greenhouse gas emissions in the first place, is to ramp up our investment in post-disaster aid and climate adaptation.

On the issue of disaster relief, this means beefing up our disaster response system and offering long-term support to the victims of climate disasters. This might mean allocating more money to FEMA's disaster-relief unit, investing in more and better-funded home buyouts, or providing more generous moving stipends to those who relocate without the benefit of an insurance payout. Beyond these basic measures, we might think more creatively about how to support people who want to move away from risky places: it would be a simple matter to offer expanded tax credits for people who start mortgages in new cities after disasters, and lawmakers could provide further incentives for those who relocate out of flood zones.

Increased investment in climate adaptation, meanwhile, would mean more initiatives like the Chesterfield Heights flood resilience project in Norfolk, more investments in forest thinning and defensible space to combat wildfires, and more support for agricultural interventions that would make farming less water-intensive and more sustainable. Since we cannot relocate everyone out of the riskiest areas, we also have an obligation to reduce the disaster risk that people in those areas face, especially in large cities where resilience efforts tend to pay for themselves many times over. The bipartisan infrastructure bill signed into law by President Biden in 2021 contained around $50 billion for resilience measures, by far the largest amount ever allocated for such projects, but even this money will not be enough to address the full scale of the risk. Solving regional challenges like the western water crisis, for instance, will require billions more to solve legal disputes between parties, invest in new water storage systems, and compensate those who lose jobs in farming and ranching. Recent research shows that these policies enjoy widespread support: a poll from the think tank Data for Progress found that 83 percent of likely voters favor more housing assistance for those displaced by climate disasters, and 85 percent of likely voters favor more funding for resilient infrastructure.

We also need to address systemic issues in the mortgage and insurance markets that control property in climate-vulnerable areas. The private fire insurance industry and the public National Flood Insurance Program are both broken and in dire need of reform, for very different reasons. Private fire insurance companies deny coverage to people who need it, even when those people take steps to lower their fire risk; the government flood insurance program, meanwhile, subsidizes many people to stay in the riskiest areas, even as it offers no discounts or grants for low-income households who cannot afford basic flood coverage. It is incumbent on Congress and state regulators to find long-term remedies for these problems. Furthermore, federal regulators must require private lenders to account for climate change risk in their portfolios, so that banks and consumers share the same information about property risks.

Furthermore, we need policy reform outside the specific areas of insurance and disaster relief. We cannot solve the issue of climate displacement without solving the issue of housing affordability overall, and there are distinct steps that local and federal governments can take to address these crises. The federal program that offers housing vouchers to low-income renters, for instance, only receives enough funding to serve a quarter of eligible Americans, which is why every large city has a "lottery" where thousands of people wait for access to affordable apartments. If Congress were to expand funding for that program so that every eligible person received a voucher, it would help ensure that renters in cities like Houston could find a new apartment if a storm destroyed their old one. It would also protect vulnerable Californians from slipping into homelessness after fire events and would prevent those same people from needing to rent in the most fire-prone areas. Going even further, the federal government could invest in high-density social housing in major cities, giving people who migrate to these places an affordable and sustainable place to stay; local governments could also relax zoning restrictions to allow for more construction, which would bring down costs.

The issue of international migration presents an even more profound challenge. The politics of immigration reform are complex, vicious, and almost impossible to navigate, and no bipartisan (or even partisan) plan seems forthcoming, but as the decades go on, the issue will get harder and harder to ignore. If the United States does not take steps to open its borders to more migrants, it will not only condemn millions of people to destitution and danger but will also miss out on an enormous opportunity for economic growth and revitalization. Despite the mounting risk of climate disaster in some places, there is no shortage of room for new arrivals.

The twin crises of housing displacement and international immigration both demand a centralized federal plan for adapting to climate change, something that has never yet existed. For the first time in more than a century, the federal government will need to take an active role in facilitating migration, and will need to spend unprecedented sums to help internal and external climate refugees find safe shelter. There are any number of ways to do this at the national and local level—land trusts for future migrants, high-ground mandates for new construction—but the first step to solving these massive crises will be to acknowledge them as crises. This is something our leaders are just beginning to do.

———

Even with an adaptation initiative of adequate scale, it will be impossible to save every place. There will be coastal bayou towns that will wash away, small riverside villages that succumb to the current, beautiful mountain towns like Paradise that will burn in raging wildfires. Countless different communities, pockets of idiosyncratic subculture, repositories of surprising and cherished history, will have to be abandoned. Thousands and thousands of homes, some historic, others just beloved by their owners, will burn up or dry out or fall into the sea. No amount of political intervention and invest-ment can stop that from happening, and thus for many Americans climate change will look like letting go of their old ideas of home, ideas that in many cases are synonymous with ideas of the American dream.

On the other side of this loss, there is an opportunity—an opportunity to rethink this idea of home, or reform it, or create it anew. Tackling cli-mate change in the short term will mean ending the hegemony of oil and gas, decarbonizing the economy, and retrofitting vulnerable communities for a dangerous new world. In the long term, it will require building a po-litical system that can support people through the climate shocks that are already inevitable. If the most pervasive impact of the climate crisis will be property destruction, it follows that the best way of resolving the crisis will be to ensure that everyone has access to housing, before and after disasters. The government cannot protect every home from destruction, but it can protect every person from the worst consequences of losing their home.

A guarantee like this sounds radical because it undermines a belief that has long been foundational to the American psyche, and has helped fuel the idea of unfettered expansion: the belief in individual responsibility. The philosophy of individualism that descended from Ralph Waldo Emerson to Horatio Alger to Milton Friedman remains foundational to the political culture of this country. One of the defining principles of American life over the past two centuries has been that each man is responsible for pursuing his own happiness and that, barring an act of God, it is his own fault if he does not attain it. The next hundred years will make those "acts of God" far more frequent, so that by the end of the century the risks of climate disaster will no longer be marginal but endemic to property ownership. The government will be the only entity capable of shouldering those risks, which means that millions of people will need to learn to rely on public assistance for shelter and safety.

The corollary to the question of responsibility, then, is the question of *right*, and in particular the question of what rights we grant to climate victims. Since the predominant consequence of the climate crisis will be the destruction of property, the main right in question is the right to shelter. In our society this right is tied to property ownership, and one's ability to recover after a disaster depends on one's material assets. In a just world, this would not be the case. Responding to future climate displacement will require us to recognize that all human beings are entitled to safe housing, not just those who have bought a house and purchased insurance to protect it.

The United Nations has already affirmed the inalienable rights of climate refugees to social and economic stability, whether they are displaced within their own countries or across national borders. The body's high court ruled in 2020 on the case of Ioane Teitiota, a man from the sinking island nation of Kiribati who had applied for asylum in New Zealand, and found that climate change refugees cannot be returned to their country of origin; its earlier compact on refugees has affirmed the rights of internal and international migrants to shelter and dignity. As with most United Nations decisions, this ruling only has force to the extent that countries choose to follow it, and so far few countries have, but the outline for a more humane framework already exists. Protecting some people from climate displacement means protecting everyone from displacement in general.

There is no doubt that the most urgent task we face as a society is to slash carbon emissions as fast as possible. If we fail to keep overall warming below two degrees Celsius, the scenarios depicted in this book will look halcyon by comparison to the reality by the end of the century. Even if we do manage to phase out fossil fuels and stop planetary warming, though, the most difficult and agonizing questions will remain unanswered. How do we adapt to the grave new world of climate change, and who should bear the burden of that adaptation? The future depends on how we choose to answer these questions.

With every passing year, indeed with every new season, climate change forces more people out of their homes. By the time you read this, some new and unforeseen calamity may have already come to dominate the headlines. The levees are already breaking, the rivers are already running dry, the fire is already snaking through the forest. The world is already being remade, but its future shape is far from set in stone. The next century may usher us into a brutal and unpredictable world, a world in which only the wealthiest and most privileged can protect themselves from dispossession, or it may usher us into a fairer world—a world where one's home may not be impregnable, but where one's right to shelter is guaranteed.

Both worlds are possible. We still have time to choose between them.

Acknowledgments

This book would not exist but for my agent, Sarah Fuentes, and my editor, Emily Simonson. It was Sarah who emailed me in the early days of the pandemic to ask whether I had ever considered writing a book, and who helped me through the proposal and everything beyond that. It was Emily who took a chance on the proposal and who brought the book to life. I am deeply indebted to both of them.

I am also indebted to the hundreds of people who agreed to speak with me during the reporting process, and especially to the disaster victims whose stories fill these pages. It is a very brave thing to sit down with a stranger and relive some of the worst moments of your life, and I will always be grateful to the people who spoke with me and expected nothing in return. It is in doing right by these people that this book will succeed or fail.

There were many people who offered me advice and guidance as I arrived in new places to gather stories for the book. I want to mention a few of these people here. Thank you to Kendall Klay and Lenore Baker in Big Pine; Eartha Mumford and Treda Berry in Kinston; Arthur Dawson and Jamie Crozat in Sonoma County; Alton Verdin, Christine Verdin, Albert Naquin, and Jake Billiot in Pointe-aux-Chenes; James Wade and Esther Del Toro in Houston; Paco Ollerton in Casa Grande; and Mason Andrews, Paige Pollard, and Karen Speights in Norfolk. Kelsea Norris, Allie Baer

Chan, Don and Mary Ellender, Rob Sperry-Fromm, Justin Fife, and Mike Galperin let me stay with them while I traveled. Kezia Setyawan, Jessie Fay Parrott, Sophie Kasakove, and Rachel Wolfe were ideal comrades during the harrowing aftermath of Hurricane Ida. I am also grateful for the work of my research assistant, Maddie Parrish, who fact-checked much of the early manuscript and who conducted interviews in Spanish as well. The work of Elizabeth Rush, Rebecca Elliott, Dan Shtob, Sonia Shah, Christopher Flavelle, Brooke Jarvis, Isabel Wilkerson, and many other writers provided invaluable inspiration and guidance.

I would also like to thank the many friends and colleagues of mine who read early drafts of these chapters and offered sage feedback. Writers always say this, but it's true: writing a book is a collaborative process, not something one does by oneself. I hope the people who helped me can see their individual wisdom reflected in the finished product.

I have benefited during my career as a journalist from many wonderful mentors and colleagues. This began at my high school newspaper, the *Steinbrenner Oracle*, which at the time was run by the ever-patient James Flaskamp. I spent many long hours working on InDesign layouts in the company of Brandon Mauriello, Kyle Dunn, and Natalie Barman. I was the only one out of the four of us who ended up as a journalist, but I still believe that any of the other three would have done a better job.

It was in Chicago, at the *South Side Weekly*, that I first began to think of myself as a reporter. That would never have happened without the mentorship of Harrison Smith, Bea Malsky, Hannah Nyhart, Osita Nwanevu, Harry Backlund, Spencer McAvoy, John Gamino, and others. I learned just as much if not more from the time I spent working alongside Maha Ahmed, Christian Belanger, Mari Cohen, Ellie Mejía, and the rest of the *Weekly*'s volunteer staff.

I would never have survived in New York City had it not been for Naomi Gordon-Loebl, the best first boss one could ask for, who taught me how to fact check at *The Nation* and made me employable after I left. Likewise, I would not have made it through four-plus years of freelancing without the friendship and support of Miguel Salazar, Jasper Craven, and Glyn Peterson, my former *Nation* colleagues. The list goes on: Ted Hart at *New York* and Corinne Iozzio at *Popular Science* took me on as a fact-checker, Jess Bergman at *The Baffler* took a chance on me and commissioned the essay that led to this book, Christopher Shay at *The Nation* launched me on the Postal Service beat, and Laura Reston at the *New Republic* made me a better writer and thinker. Jamie Keiles rekindled in me the curiosity that makes for great journalism. Reeves Wiedeman taught me more about reporting in three months than I had learned in the previous three years. John Thomason, Zoya Teirstein, and Nikhil Swaminathan at *Grist* helped me move on to the climate beat full-time.

I am blessed to have in New York a group of friends with whom I can discuss Henry James and LeBron James in equal measure. I am grateful to Kay Li, Brad Cohn, Adam Gluck, Rohan Sharma, Konje Machini, Gaby del Valle, Sam Raskin, Noah Goldberg, Ezra Marcus, Charlie Dulik, Jordan Larson, Amanda Arnold, Marian Bull, Matt Gasda, and Billy Lennon for their companionship. Thank you to Matt Stieb and Chris Crowley, who are always ready to share an appletini, and to the ever-welcoming James Walsh and Carina Guiterman. Thank you to the editors of *The Drift*—Rebecca Panovka, Krithika Varagur, Sophie Haigney, Elena Saavedra Buckley, and others—for breathing new air into my life in New York.

The University of Chicago was far from the easiest place to spend one's college years, but it was during those years that I made the best friends I think I will ever have. Luke Sironski-White, Asher Baumrin, Kevin Otradovec, and Michael Borde know that I owe them more than I will ever have the words to express. Andrew Eckholm and Jackson Roth came up to me while I was crying over a breakup on a park bench and changed my life. I am thankful beyond measure for the friendship of Daly Arnett, Clark Halpern, Cathryn Jijón, Julianna St. Onge, Freddy Davis, Terry Hines, Adonia Bekele, Lucia Ahrensdorf, Mary Otoo, Andrew Fialkowski, Nick

Hahn, Emma Herman, Emily Lipstein, Molly Robinson, Ang Zhang, and many others. And everything I do, even now, is for Myles Johnson and CJ Lakoduk.

An adolescence in Lutz, Florida, was made only just tolerable by my friends there. Austin Schmitz, Quinn Sahler, Ethan Huber, Brandon Nguyen, Elliott Smith, Dylan Carlson, Nick Novy, Jonathan Fuentes, Jeffery Ccsari, Jackie Braje, Katie Shane, Natalie Barman, Ryan Bianchet, and Morgan Melatti Bianchet all helped make me the person I am. I still have a large piece of all of them with me today, and that is a great comfort.

I would not be a writer if it were not for my parents, Kevin and Rebecca Bittle, who never questioned my ambition to be a writer or my ill-conceived plan to freelance in New York. When they took me to Barnes & Noble or printed off the amateur novels I wrote in my spare time, they were making this book possible. None of it would have happened without them, or without the constant and unwavering support of my entire family, in particular my grandparents Leon and Judith Major and my cousin Natasha Major. I am lucky as well to have had my great-uncle Mark Strand as a model of writerly grace.

My last and most profound gratitude is to Sara Bosworth, who made all this possible. If it were not for you, Sara, I would never have tried.

Notes

INTRODUCTION

xiii *"because we knew everybody"*: Brittny Mejia, " 'We've Lost Everything': Survivors Reeling as Downtown Greenville Leveled by Dixie Fire," *Los Angeles Times*, August 5, 2021.

xiii *"rednecks, ranchers, [and] cowboys"*: Margaret Elysia Garcia, "Eulogy for Greenville, My Beautiful Hometown Lost to Wildfire," *The Guardian*, August 7, 2021.

xiv *dissolved into yellow air*: "The Dixie Fire Has Destroyed Most of a Historic Northern California Town," NPR, August 5, 2021.

xiv *apartments they could afford*: Jane Braxton Little, "17 Months Later, Dixie Fire Survivors Are Experiencing a New Kind of Disaster," *Yes!*, January 7, 2022.

xiv *insurance coverage to rebuild*: Cat Wise and Leah Nagy, "Many Greenville Residents Struggled to Get Fire Insurance. Then the Dixie Fire Came," *PBS NewsHour*, October 11, 2021.

xiv *slipping into new lives*: Dani Anguiano, "Revisiting Greenville: The Mountain Town Destroyed by California's Largest Wildfire," *The Guardian*, January 5, 2022.

xiv *frozen in the moments after the fire*: Alex Wigglesworth, "Greenville Was Destroyed by Wildfire. Can It Be Rebuilt to Survive the Next One?," *Los Angeles Times*, May 30, 2022.

xv *historic fire near Lake Tahoe*: Neil Vigdor and Thomas Fuller, "Evacuations Ordered Near Lake Tahoe as the Caldor Fire Chokes Region," *New York Times*, August 30, 2021.

xv *monstrous hurricane in Louisiana*: Jake Bittle, "Hurricane Ida Is a Manmade Disaster," *The New Republic*, August 30, 2021.

xv *drought across the Southwest*: Denise Chow, "U.S. Megadrought Worst in at Least 1,200 Years, Researchers Say," NBC News, February 15, 2022.

xv *weather disaster of some kind*: Sarah Kaplan and Andrew Ba Tran, "Nearly 1 in 3 Americans Experienced a Weather Disaster this Summer," *Washington Post*, September 4, 2021.

xvi *cities like New York and Chicago*: Isabel Wilkerson, *The Warmth of Other Suns: The Epic Story of America's Great Migration* (New York: Vintage, 2011), 8–17, 165–79, 183–223.

xvi *the former plantation states*: Richard M. Mizelle Jr., *Backwater Blues: The Mississippi Flood of 1927 in the African American Imagination* (Minneapolis: University of Minnesota Press, 2014).

xvi *this seismic shift*: Isabel Wilkerson, "The Long-Lasting Legacy of the Great Migration," *Smithsonian*, September 2016.

xvi *uprooting millions of people*: Abrahm Lustgarten, "How Climate Migration Will Reshape America," *New York Times*, September 15, 2020. It is impossible to project with certainty how many people climate change will displace, but even conservative estimates suggest the totals will eclipse the Great Migration. Lustgarten's data analysis in the cited article suggests that by the end of the century at least four million people in the United States will live "in places decidedly outside the ideal niche for human life," and that's not counting the tens of millions of people who live at risk of occasional disasters like hurricanes and wildfires. Indeed, more than six million people in the United States lost their homes to climate disasters between 2016 and 2020. Most of those people did not relocate, but if even a small proportion of displaced persons move every year, the total number of movers will easily surpass six million by the end of the century.

xviii *killing hundreds of people*: Nadja Popovich and Winston Choi-Schagrin, "Hidden Toll of the Northwest Heat Wave: Hundreds of Extra Deaths," *New York Times*, August 11, 2021.

xviii *the storm itself did in Louisiana*: Bobby Caina Calvan, David Porter, and Jennifer Peltz, "More than 45 Dead After Ida's Remnants Blindside Northeast," Associated Press, September 2, 2021.

xviii *deep freeze in Texas*: Erin Douglas, "Winters Get Warmer with Climate Change. So What Explains Texas' Cold Snap in 2021?" *Texas Tribune*, December 14, 2021.

xviii *tornado squall in Kentucky*: Ryan Van Velzer, "Kentucky Faces Greater Tornado Risks Because of Climate Change," 89.3 WFPL News Louisville, December 16, 2021.

xviii *in the Denver suburbs*: David Wallace-Wells, "The Return of the Urban Firestorm," *New York*, January 1, 2022.

xx *"life on the move"*: Sonia Shah, *The Next Great Migration: The Beauty and Terror of Life on the Move* (London: Bloomsbury, 2020).

CHAPTER 1: THE END OF THE EARTH

3 *property of a local hermit*: Interview with Jen DeMaria and Harry Appel, November 2020.

4 *fruit grove on Big Pine*: Interview with Patrick Garvey, November 2020.

4 *build a global nursery*: Patrick Garvey, "Grimal Grove: The Story," Grimal Grove, http://www.grimalgrove.com/The-Story.

4 *nowhere else in the continental United States*: Gretchen Schmidt, "Digging Up the Past: Grimal Grove," Edible South Florida, June 13, 2016.

4 *ahead of Labor Day weekend*: John Cangialosi et al., "National Hurricane Center Tropical Cyclone Report: Hurricane Irma," National Hurricane Center, September 24, 2021.

4 *while still in the Atlantic*: Cangialosi et al., "Hurricane Irma." This was the second-longest Category 5 episode on record. The storm also had the longest sustained episode of 185-mile-per-hour winds, maintaining those winds for thirty-seven hours.

5 *both sides of the highway*: National Weather Service, "Hurricane Irma Local Report/Summary," US Department of Commerce, NOAA, https://www.weather.gov/mfl/hurricaneirma.

5 *more than 130 miles per hour*: Cangialosi et al., "Hurricane Irma."

5 *if the worst came to pass*: Christopher Mele, "How to Get People to Evacuate? Try Fear," *New York Times*, October 6, 2016; and William South, "I'm a Meteorologist. I'm Staying Behind in the Florida Keys to Help Save Lives," *Washington Post*, September 9, 2017. Also interviews with Lenore Baker and Paul Kapsalis.

6 *for what seemed like hours*: Interviews with Jen DeMaria, Patrick Garvey, and "Captain" Jack Warner, November 2020.

7 *storm surge and extreme wind*: Matthew Cappucci, "Capital Weather Gang: Dissecting the Parts of a Hurricane," *Washington Post*, September 10, 2017.

9 *a haze of mosquitoes*: Interviews with Patrick Garvey, 2020 through 2022.

9 *elevating it above flood level*: "Unit 8: Substantial Improvement and Substantial Damage," National Flood Insurance Program Study Guide, FEMA, https://www.fema.gov/pdf/floodplain/nfip_sg_unit_8.pdf.

9 *turned many survivors into squatters*: Alex Harris, "It Was Hard to Find Cheap Housing in the Keys Before Irma. Now, There's 'Nothing,'" *Miami Herald*, May 2, 2018.

11 *smoldering for more than a week*: Gwen Filosa, "Whipped by a Hurricane and Burned by a Wildfire. How Did These People Survive?" *Miami Herald*, May 30, 2018.

13 *stay together longer*: Thomas Knutson et al., "Climate Change Is Probably Increasing the Intensity of Tropical Cyclones," ScienceBrief News, March 31, 2021.

13 *as it barreled toward the Keys*: Jason Samenow, "Because of Climate Change, Hurricanes Are Raining Harder and May Be Growing Stronger More Quickly," *Washington Post*, May 8, 2018. As waters in the Gulf of Mexico grow warmer, episodes of rapid intensification are much more likely.

13 *more storms like Irma*: Wayne Drash, "Yes, Climate Change Made Harvey and Irma Worse," CNN Health, September 19, 2017.

13 *for months at a time*: Patricia Mazzei, "82 Days Underwater: The Tide Is High, but They're Holding On," *New York Times*, November 24, 2019.

13 *the nation's next great port city*: Jefferson B. Browne, "The Old and the New," excerpted in "Key West: History and Sketches," Jefferson B. Browne, *The Old and the New*, published 1912 and anthologized in "Key West: History and Sketches," a University of South Florida archival project, https://fcit.usf.edu/florida/docs/k/keys03.htm.

14 *an exodus toward Miami*: Jerry Wilkinson, "History of Wrecking," Florida Keys History Museum, http://www.keyshistory.org/IK-wrecking.htm.

14 *a highway down the archipelago*: Willie Drye, *Storm of the Century: The Labor Day Hurricane of 1935* (Washington, DC: National Geographic Society, 2002).

14 *could park their boats*: "Water Quality Protection Program—Canal Demonstration Projects," presentation prepared by Rhonda Haag for the Florida Keys National Marine Sanctuary Advisory Council, December 8, 2015.

15 *from something or toward something*: Interviews with Lenore Baker and Debby Zutant, November 2020.

15 *the neighborhoods they inhabit*: Jerry Wilkinson, "History of Big Pine Key," Florida Keys History Museum, http://www.keyshistory.org/bigpinekey.html.

15 *New York City levels*: November 2021 rent data courtesy of real estate website Zumper.

16 *as fast as was necessary*: "2017 Hurricane Season: FEMA After-Action Report," FEMA, July 12, 2018, https://www.fema.gov/sites/default/files/2020-08/fema_hurricane-season-after-action-report_2017.pdf; and interview with Holly Raschein.

17 *their relief applications*: Interviews with Lenore Baker, Kendall Klay, Chris Todd, Hervé Thomas, and Patrick Garvey, November and December 2020.

17 *they drew hundreds more volunteers*: Interview with Lenore Baker, November 2020.

18 *first business on the island to reopen*: Interview with Debby Zutant, November 2020.

18 *dread about the island's future*: Interviews with Debby Zutant, Kendall Klay, Jen DeMaria, and Lenore Baker, November 2020.

19 *felt much larger*: Nancy Klingener, "Florida Keys Cope with Suicide Spike After Hurricane Irma," WLRN, July 29, 2018.

20 *fifty years in the Keys*: Interview with Connie Faast, December 2020.

21 *a quarter of the Keys's housing stock*: Valerie Bauerlein, Scott Calvert, and Jon Kamp, "Hurricane Irma Destroyed 25% of Homes in Florida Keys, FEMA Estimates," *Wall Street Journal*, September 12, 2017.

22 *outweighed the benefits*: "Florida Keys Population Dropped After Hurricane Irma, Survey Shows," Associated Press, October 22, 2018.

22 *from being rebuilt*: Alan Gomez, "Hurricane Irma Shuts the Door on Keys' Most Affordable Housing—RVs and Trailer Parks," *USA Today*, September 25, 2017.

22 *hadn't been able to afford insurance*: Interviews with Kendall Klay, Lenore Baker, Holly Raschein, and Debra Maconaughey, November 2020.

22 *wealthy homeowners and part-time vacationers*: Angelica LaVito, "Irma May Have Caused $42.5 Billion to $65 Billion in Property Damage, Report Says," CNBC, September 19, 2017. CoreLogic estimated that around 80 percent of damage was not covered by flood insurance.

22 *the county could issue*: The law restricting the county's housing supply is known as the "rate of growth ordinance," or ROGO. Its purpose is to ensure that the entire population of the county could drive to the mainland via the Overseas Highway in the span of twenty-four hours. Even before the hurricane, the county had already announced that it would not issue any new building permits after 2026, although there has been talk of revising this rule.

22 *in a single day*: Charlotte Twine, "Monroe County Only Has 2 Affordable Housing Building Permits Left," *Keys Weekly*, July 27, 2021.

23 *defining challenge of the next few years*: Interview with Debra Maconaughey, November 2020.

23 *a handful of new cottages*: Gwen Filosa, "'Keys Cottages' Are Arriving in a Working-Class Neighborhood Hit Hard by Irma," *Miami Herald*, July 3, 2018.

23 *repaired only two homes*: Sara Matthis, "Two Years Later: Rebuild Florida Program Has Only Touched Two Irma-Damaged Homes," *Keys Weekly*, September 15, 2020.

23 *since the day the storm hit*: Interview with Debra Maconaughey; and Kirk Petersen, "Small Church Meets Big Post-Hurricane Housing Need," Living Church, December 12, 2019.

24 *through the on-season rush*: Interviews with Debra Maconaughey, Chris Todd, Hervé Thomas, and Holly Raschein, November and December 2020.

24 *for minimum wage*: David Goodhue, "Workers Ride This Bus to Better-Paying Jobs in Paradise. They're a Long Way from Home," *Miami Herald*, September 20, 2019.

26 *massive ocean disturbances*: Eugene A. Shinn and Barbara H. Lidz, *Geology of the Florida Keys* (Gainesville: University Press of Florida, 2018).

26 *the end of this century*: NOAA's Sea Level Rise Viewer mapping tool shows that five feet of sea-level rise would submerge every major island in the Keys, except for Key Largo and a few sections of Islamorada and Key West. Even then, most of the dredged subdivisions of Key Largo and Islamorada would be underwater.

26 *somewhere safer*: Prashant Gopal, "America's Great Climate Exodus Is Starting in the Florida Keys," *Bloomberg Businessweek*, September 20, 2019.

27 *study, deliberation, decision*: Interview with Rhonda Haag, December 2020.

27 *houses on the road*: Christopher Flavelle and Patricia Mazzei, "Florida Keys Deliver a Hard Message: As Seas Rise, Some Places Can't Be Saved," *New York Times*, December 4, 2019.

27 *through the year 2045*: "Countywide Roadway Vulnerability Analysis and Capital Plan," produced by Rhonda Haag for Monroe County, provided to the author by Rhonda Haag.

28 *the money that would be needed to protect them*: Alex Harris, "New South Florida Climate Change Financial Report: Spend Billions or Lose Much, Much More," *Miami Herald*, October 13, 2020.

28 *the worst months of flooding*: Interviews with Rhonda Haag, 2020 and 2021.

29 *drive out to the main highway*: David Goodhue, "Last Year, This Neighborhood Flooded for 92 Days. Residents Fear They're in for a Replay," *Miami Herald*, October 3, 2020.

29 *three or four inches of khaki-colored water*: "Flood Control Barriers Tested in Key Largo's Twin Lakes," *Keys Weekly*, October 15, 2020.

29 *first to get its roads raised*: Interview with Andy Sikora, a longtime resident of Stillwright Point, December 2020.

30 *at the end of the summer*: Interview with Patrick Garvey in November 2020.

CHAPTER 2: AFTER THE FLOOD

35 *pulsing, ponderous silence*: Author's own observations from site visit in January 2021.

36 *teachers and social workers*: "Historic and Architectural Resources of Kinston, North Carolina," National Register of Historic Places Multiple Property Documentation Form, September 25, 1989, https://npgallery.nps.gov/GetAsset/d15f8054-b35f-4b7e-99c2-24e26df06e3b.

36 *an almost unaccountable blessing*: Interviews with William Lawson, Treda Berry, Eartha Mumford, and Iris Silvers, January and February 2021.

37 *where else would you want to go?*: Interviews with Iris Silvers and Paul Simmons, January 2021.

38 *work the same cotton fields*: Eric Foner, *A Short History of Reconstruction*, abridged ed. (New York: Harper Perennial Modern Classics, 2015).

38 *almost entirely dependent on their unpaid labor*: "Appendix H: Princeville White Paper," from "Princeville, North Carolina Flood Risk Management Integrated Feasibility Report and Environmental Assessment," U.S. Army Corps of Engineers, December 2015. Report contains historical survey of post-Emancipation conditions in eastern North Carolina.

38 *to work on the docks*: Daniel H. de Vries, "Temporal Vulnerability: Historical Ecologies of Monitoring, Memory, and Meaning in Changing United States Floodplain Land-scapes," University of North Carolina at Chapel Hill, May 2008, 75, and see note 2.

39 *William and his six siblings*: Interviews with William Lawson Sr. and William Lawson Jr., February 2021.

39 *"go[ing] to higher ground"*: De Vries, "Temporal Vulnerability," 75.

39 *jobs in the area for Black people*: Interview with William Lawson Sr., February 2021.

39 *from around the state*: Ted Sampley, "Mayor's Unprecedented Attempt to Clean Up Sugar Hill 'Vice District' Failed," *Olde Kinston Gazette*, August 1998.

40 *many workers from Lincoln City*: Lu Ann Jones, "DuPont Comes to Tobacco Road: Oral History and Rural Industrialisation in the Post–Second World War American South," *Oral History* 42, no. 1 (Spring 2014): 35–46.

40 *like the Ohio River*: The company reached a $16.5 million settlement with the Environmental Protection Agency in 2005 over allegations that it allowed the chemical PFOA to leak into West Virginia waterways, and paid out $27 million in a separate 2021 settlement over similar allegations that it dumped PFOA and PFAS into Ohio drinking water.

40 *Kinston and other cities*: Cullen Browder, "NC Ignored for Decades Ways to Limit Neuse River Flooding," WRAL, November 13, 2018.

40 *the second dam, near Kinston*: De Vries, "Temporal Vulnerability," 105–6; and interview with Lim Vallianos, a retired engineer who worked in the US Army Corps of Engineers Wilmington District.

40 *new reservoirs in the area*: Cullen Browder, "NC Ignored for Decades Ways to Limit Neuse River Flooding," WRAL, November 13, 2018.

41 *from all over the region*: "Kinston Places Put 'Off Limits'—Marine Corps Authorities Act to Stamp Out Diseases," Raleigh *News and Observer*, September 6, 1944, and "Kinston Homes Evacuated," Raleigh *News and Observer*, February 20, 1960.

41 *cash stipend to move somewhere else*: "Salvaging a Business and the Environment," Raleigh *News and Observer*, November 7, 1999; and Monica McCann, "Case Study of

Floodplain Acquisition/Relocation Project in Kinston, NC," University of North Carolina at Chapel Hill, 2006.

42 *made the math a lot easier*: Interview with Roger Dail, January 2021.

42 *spread more money around*: Interview with Orrin Pilkey, December 2020; and Gilbert Gaul, *The Geography of Risk: Epic Storms, Rising Seas, and the Cost of America's Coasts* (New York: Sarah Crichton Books, 2019), 11–47, 103–66. I am indebted to both Mr. Pilkey and Mr. Gaul for their decades of work on the issue of development in floodplains and shorelines.

43 *before they needed to be repaired or replaced*: Interview with Orrin Pilkey, December 2020; and Orrin H. Pilkey and Katharine L. Dixon, *The Corps and the Shore* (Washington, DC: Island Press, 1996), 120–25.

43 *destroyed again a few years later*: Interview with A. R. Siders, March 2021.

44 *Or so the theory went*: Congressional Research Service, *Introduction to the National Flood Insurance Program (NFIP)*, January 2021.

44 *even at subsidized prices*: National Research Council, *Affordability of National Flood Insurance Program Premiums: Report 1* (Washington, DC: National Academies Press, 2015).

44 *preventing future disaster losses*: Congressional Research Service, *FEMA's Hazard Mitigation Grant Program: Overview and Issues*, March 25, 2009.

44 *what had become a permanent problem*: Missouri State Emergency Management Agency, *Stemming the Tide of Flood Losses: Stories of Success from the History of Missouri's Flood Mitigation Program*, undated, circa 2000.

45 *the only option*: Interview with Roger Dail, January 2021.

46 *as long as they could*: Interviews with Roger Dail, Eartha Mumford, and Treda Berry, January and February 2021.

46 *would have nowhere to go*: Jason Hardin, "Floyd Puts Fear into Local Officials," *Kinston Free Press*, September 15, 1999. Accessed in the microfilm archives at Lenoir Community College in Kinston.

46 *on the phone in his office*: Jennifer Shrader, "Neuse Expected to Crest Sunday," *Kinston Free Press*, September 11, 1999.

46 *Dail told the paper*: Jennfier Shrader, "Eye Expected at or Near Kinston. Neuse Expected to Reach 24 Feet Next Week," *Kinston Free Press*, September 16, 1999.

46 *fifteen inches of rain*: Hurricane Floyd special issue, *Kinston Free Press*, September 17, 1999.

47 *a few inches an hour*: Bonnie Edwards, "Crest Level Becomes Guessing Game Again," *Kinston Free Press*, September 20, 1999.

47 *leaking fecal matter into the river*: McCann, "Case Study of Floodplain Acquisition /Relocation Project in Kinston, NC."

47 *drainage pipes and sewage lines*: Jennifer Shrader, "Some Residents Allowed to Return Home," *Kinston Free Press*, September 26, 1999.

47 *runoff from factories*: David Herring, "Hurricane Floyd's Lasting Legacy: Assessing the Storm's Impact on the Carolina Coast," NASA Earth Observatory, March 1, 2000.

47 *along with the current*: Kirk Ross and Darryl Fears, "Flooded North Carolina Farms Are Likely Littered with Drowned Livestock," *Washington Post*, October 11, 2016.

47 *to avoid vomiting*: "Hurricane Florence Could Flood North Carolina's Hog Manure Pits, Taint Drinking Water," Associated Press, September 12, 2018.

48 *"we really didn't have to do that much"*: Interview with Roger Dail, January 2021.

50 *made up of people, not buildings*: Interviews with Eartha Mumford, Elwanda Ingram, Treda Berry, and Paul Simmons, January through March 2021.

52 THE LEGEND HAS RETIRED: Fran Daniel, "Elwanda Ingram Says Goodbye to WSSU After 37 Years," *Winston-Salem Journal*, December 25, 2016.

52 *tended by their mother*: Interview with Elwanda Ingram, February 2021.

53 *"'You've got to go'"*: Interview with Elwanda Ingram, February 2021.

54 *restore their eroded beaches*: A. R. Siders and Jesse Keenan, "Variables Shaping Coastal Adaptation Decisions to Armor, Nourish, and Retreat in North Carolina," *Ocean & Coastal Management* 183 (October 2019): 105023.

54 *first in line to be sacrificed*: Interview with A. R. Siders, March 2021.

54 *she told me*: Interview with Elwanda Ingram, February 2021.

56 *T-shirt factory was on its way out*: Mid-Atlantic Associates, "Phase I Environmental Site Assessment: Former Kinston Shirt Factory," prepared for Adam Short, Kinston city planner, on July 27, 2015.

56 *plant outside the city*: Adam Wagner, "In Eastern North Carolina, Matthew 'Worse Than Floyd,'" Associated Press, October 15, 2016.

56 *new home inside the city limits*: Interviews with Roger Dail and Tony Sears, city manager of Kinston, January and February 2021.

56 *open only to whites*: Interview with Joseph Tyson, Kinston city council member, February 2021.

57 *"other communities facing similar situations"*: "Out of Harm's Way: Relocation Strategies to Reduce Flood Risk," NOAA Office of Coastal Management Peer-to-Peer Case Study. The planner quoted in the article, Adam Short, declined my request for an interview.

57 *"a lot of bitterness"*: De Vries, "Temporal Vulnerability," 71.

58 *higher property taxes and steep electric bills*: Interviews with Joseph Tyson, William Lawson Jr., and Eartha Mumford, January and February 2021.

58 *households that entered foreclosure*: Interviews with Roger Dail and Tony Sears, January and February 2021.

58 *moving across the country*: Interviews with Eartha Mumford, Paul Simmons, and El-wanda Ingram, January and February 2021.

59 *down the street from everyone else*: Interview with Eartha Mumford, January 2021.

60 *swelled to three thousand*: Mike Valerio, "Ghost Town Reels from Floyd, 13 Years Later," ABC 12 News, September 18, 2012.

61 *across the state and beyond*: Aaron Deane, "The 'Lincoln City' Community Recon-nects in Kinston," WNCT9, May 26, 2019. Also interviews with Tony Sears, Eartha Mumford, and Paul Simmons.

62 *that once existed there*: Elyse Zavar, "An Analysis of Floodplain Buyout Memorials: Four Examples from Central U.S. Floods of 1993–1998," *GeoJournal* 84, no. 3 (February 2019): 135–46.

63 *sent donations from afar*: Front pages of the *Tarboro Southerner*, September 15–October 5, 1999. Accessed in the microfilm archives of the Edgecombe County Memorial Library in Rocky Mount, North Carolina.

64 *voted three to two to rebuild*: Interviews with Calvin Adkins, longtime resident of Princeville, and Delia Perkins, former mayor of Princeville, January 2021.

64 *jeopardize the town for years to come*: US Army Corps of Engineers, *Princeville, North Carolina Flood Risk Management Integrated Feasibility Report and Environmental Assessment*, December 2015.

64 *the plan came too late*: Interview with Bobbie Jones, current mayor of Princeville, January 2021.

65 *finagled FEMA buyouts on their own*: Interviews with Bobbie Jones and Calvin Ad-kins, January 2021.

65 *along lines of race and class*: Amanda Martin, *Race, Place, and Resilience: Spatial Equity in North Carolina's Post-Disaster Buyout Program*, University of North Carolina at Chapel Hill, 2019. I am indebted to Dr. Martin for her exceptional work on this thesis and for her article "After the Flood, the Decision to Re-build or Leave Permanently," published on the website of the *Carolina Planning Journal*.

CHAPTER 3: BURNOUT

69 *cool expanse of the Pacific*: Anne Mulkern, "In California Fires, a Starring Role for the Wicked Wind of the West," E&E News, December 6, 2017.

69 *to a Category 1 hurricane*: Clifford F. Mass and David Owens, "The Northern Califor-

nia Wildfires of 8–9 October 2017: The Role of a Major Downslope Wind Event," *Bulletin of the American Meteorological Society* 100 (February 1, 2019).

69 *potential fires in the area*: John Martinez et al., *Tubbs Fire Investigation Report*, Cal Fire, January 24, 2019, 3.

70 *inevitable that something would ignite*: Steve Gorman, "Probe Finds PG&E Power Lines Sparked Deadly 2017 California Wildfires," Reuters, June 8, 2018.

70 *sparks onto the surrounding vegetation*: Martinez et al., *Tubbs Fire Investigation Report*, 27–30.

70 *the ridge behind the house*: Anne Ward Ernst, "Family Sees 'Wall of Orange' Near Their Home at Start of Tubbs Fire," *Napa Valley Register*, November 1, 2017.

70 *called for reinforcements*: Martinez et al., *Tubbs Fire Investigation Report*, 12.

70 *drying out the new vegetation into kindling*: Interview with Arthur Dawson, fire ecologist and forestry consultant who has studied historical fire ecology in Sonoma County. See also Mark Bove, "Rain Fuels Wildfire Risk," Munich RE, April 4, 2018.

71 *local or state firefighters to control*: Troy Griggs et al., "Minutes to Escape: How One California Wildfire Damaged So Much So Quickly," *New York Times*, October 12, 2017.

71 *hampering the progress of any rogue blaze*: Interview with Vicki and Mark Carrino, September 2021.

72 *charred outline of the home's foundation*: Photographs and videos provided by Vicki and Mark Carrino.

73 *scattering livestock down the hills*: Interview with Vicki and Mark Carrino, corroborating accounts provided by Katie Bower, Lisa Mattson, and Jamie Crozat, September and October 2021.

74 *complex of courtyard apartments*: Interviews with Trent Yaconelli, Gabe Heskett, José Guzman, and other residents of Coffey Park, September 2021; numerous contemporary news accounts and photographs from after the scene show the path of fire damage in this area.

74 *weren't even tracking the blaze*: Interview with Kevin Tran, September 2021.

75 *just a mile away*: Interview with Trent Yaconelli, resident of Coffey Park, September 2021.

77 *the world's fifth-largest economy*: United States Census Bureau data, 1960 and 2010 censuses.

78 *dominate the national economy*: For one of the best contemporary histories of California, see Miriam Pawel, *The Browns of California* (London: Bloomsbury Press, 2018).

78 *bring low-income people into their midst*: Conor Dougherty, *Golden Gates: The Housing Crisis and a Reckoning for the American Dream* (New York: Penguin, 2020), 1–15, 63–93.

78 *"hadn't been greased enough"*: Dougherty, *Golden Gates*, 3.

79   *incentives to build new housing*: F. V. Ferreira, "You Can Take It with You: Proposition 13 Tax Benefits, Residential Mobility, and Willingness to Pay for Housing Amenities," *Journal of Public Economics* 94, nos. 9–10 (October 2010): 661–73.

79   *they had no reason to build*: Matt Levin, "Too Few Homes: Is Prop. 13 to Blame for the State's Housing Shortage?," KPBS/Calmatters, October 30, 2018.

79   *left it open to abuse*: M. Nolan Gray, "How Californians Are Weaponizing Environmental Law," *The Atlantic*, March 12, 2021.

79   *real motivations were less charitable*: Jennifer Hernandez, "California Environmental Quality Act Lawsuits and California's Housing Crisis," *Hastings Environmental Law Journal* 24, no. 1 (Winter 2018): 21–71.

79   *do not have one*: "California's High Housing Costs: Causes and Consequences," California Legislative Analyst's Office, 2015, https://lao.ca.gov/reports/2015/finance/housing-costs/housing-costs.pdf.

80   *boom in wine consumption*: "Complete Napa Valley California Wine History from Early 1800s to Today," *Wine Cellar Insider*, accessed October 2021, https://www.thewinecellarinsider.com/california-wine/california-wine-history-from-early-plantings-in-1800s-to-today/.

80   *for tours and tastings*: "History of Wine in the Napa Valley," Napa Valley Vintners, accessed October 2021, https://napavintners.com/napa_valley/history.asp/.

80   *places to stay as well*: Interviews with Stu Smith, co-owner of the Smith-Madrone winery, and Ryan Klobas, CEO of the Napa County Farm Bureau, July 2021.

81   *one of the lowest rates in the country*: "Comprehensive Housing Market Analysis: Santa Rosa, California," US Department of Housing and Urban Development Office of Policy Development and Research, April 1, 2017.

81   *over the same period*: "Sonoma County's Affordable Housing Catastrophe," *Sonoma Valley Sun*, July 7, 2018.

81   *meager supply of shelter beds*: Angela Hart, "Sonoma County Supervisors to Consider Housing Homeless in Tiny Homes," Santa Rosa *Press Democrat*, November 1, 2015.

81   *minuscule number of available homes*: Kirk Johnson and Conor Dougherty, "California Fires Leave Many Homeless Where Housing Was Already Scarce," *New York Times*, October 15, 2017.

81   *a state of outright turmoil*: Interview with Belén Lopez-Grady, deputy director at North Bay Organizing Project, August 2021. Belén and her colleagues provided tenant services and advocacy for numerous tenants and agricultural laborers who were displaced after the fire, including those who had been doubled-up in Coffey Park. I am grateful to her for the background information and guidance about the area's housing crisis.

81 *30 percent overnight*: Interviews with Katie Bower, Natalie Manning, Lisa Mattson, and Jamie Crozat, September 2021.

82 *in a few years after the recovery*: Interview with Joshua Weil, July 2021.

82 *"additional living expenses"*: Interviews with Kevin Tran, Henry Arriaga, and Jamie Crozat, September 2021.

82 *that much or even more*: "Six Santa Rosa Landlords Face Prosecution for Price Gouging During North Bay Fires Emergency," *Bay City News*, December 9, 2017.

83 *second homes they had been leasing out*: Interviews with Belén Lopez-Grady, Kevin Tran, and Hung Tang, resident of Coffey Park, August and September 2021.

83 *another five years while they saved*: Interviews with Henry Arriaga, September and October 2021.

86 *the fire-prone hills*: Interview with José Guzman, September 2021.

89 *for suitable building sites*: Damian Bacich, "Fr. José Altimira and Mission San Francisco Solano," California Frontier Project, February 23, 2018.

89 *"The place is bare of thick woods"*: Arthur Dawson, "Sonoma Valley: The Oral History Project," Baseline Consulting, compiled between 2002 and 2021. Project is ongoing.

90 *forests remained healthy nonetheless*: Interview with Arthur Dawson, September 2021.

90 *Smokey Bear*: "U.S. Forest Service Fire Suppression," Forest History Society, accessed October 2021, https://foresthistory.org/research-explore/us-forest-service-history/policy-and-law/fire-u-s-forest-service/u-s-forest-service-fire-suppression/.

90 *never caught up with that research*: "The Ecological Benefits of Fire," *National Geographic* Resource Library, January 15, 2020.

90 *overlapped with wild forest*: In 1923, when Coffey Park was a small village called Boyes Springs, a fire tore down the mountains on the strength of "wind so strong it was hard to stand"; by the time it burned out, only two houses in the village were left standing. Thirty years later, the area was the site of a regional airport, and again it burned to the ground, this time because a group of airport employees decided to conduct a weed-burning operation in hundred-degree heat. See Jeff Elliott, "The Forgotten Fires of Fountaingrove and Coffey Park," Santa Rosa History, October 22, 2017.

91 *a few vineyards and farms*: Jeff Elliott, "The 1964 Hanly Fire," Santa Rosa History, September 11, 2019.

91 *the city chose growth over safety*: Interview with Arthur Dawson, September 2021.

91 *on sheer rock cliffs*: Judy Richter, "No Slowing in Santa Rosa Home-Building Boom," *San Francisco Chronicle*, October 14, 2001.

93 *to satisfy lenders and homeowners*: Interviews with Lisa and Damon Mattson, Joshua Weil, and Jamie Crozat, August and September 2021.

94 *financial incentive to stay put*: Interviews with Hung Tang, Trent Yaconelli, José Guzman, and Kevin Tran, September 2021.

95 *a few towns over*: Interview with Kevin Tran, September 2021.

95 *made up the difference*: Interview with Vicki and Mark Carrino, September 2021.

97 *has never been torn down*: Author's own impressions, from visit to Fountaingrove in September 2021.

98 *Santa Rosa adopted after the fire*: Dale Kasler and Ryan Sabalow, "Burned-Out California Town Ignores Stricter Building Codes, Even with Wildfire Threat," *Sacramento Bee*, November 15, 2019.

98 *hardened concrete walls and fireproof windows*: Interview with Jesse Oswald, chief building official for the city of Santa Rosa, July 2021.

98 *twenty-five years of cumulative underwriting profits*: Leslie Kaufman and Eric Roston, "Wildfires Are Close to Torching the Insurance Industry in California," *Bloomberg Green*, October 11, 2021.

99 *$8,000 a year*: Interview with Lisa and Damon Mattson, September 2021.

99 *drought-tolerant plants*: Meg McConahey, "Colorful Succulents Bring Beauty to Fountaingrove Garden Rebuilt After Tubbs Fire," *Sonoma Magazine*, July 2021.

100 *what the Mattsons' home was worth*: Interview with Lisa and Damon Mattson, September 2021.

100 *stop issuing new policies*: Interviews with Lisa Mattson, Vicki Carrino, and Jamie Crozat, September 2021.

101 *dropped customers in the Central Valley*: Nina Lozano, "Californians Risk Losing Insurance, Face Steep Premium Increase as Wildfires Worsen," KSBY, August 17, 2021. See also Marie Edinger, "Insurance Companies Dropping Mountain Community Homeowners Left and Right," KMPH, August 5, 2021.

101 *heading into another fire season*: Jake Bittle, "As Wildfires Worsen, More California Farms Are Deemed Too Risky to Insure," Grist, July 28, 2021.

101 *swoop in to demand collateral*: Interview with Stu Smith of Smith-Madrone, July 2021.

102 *Tubbs victims*: J. D. Morris, "On Anniversary of Tubbs Fire, PG&E Settlement Shapes Sonoma County Debate on Future," *San Francisco Chronicle*, October 7, 2020.

CHAPTER 4: THE STORY OF THE VERDINS

105 *building forks and fans of fertile land*: John Snead, Richard P. McCulloh, and Paul V. Heinrich, *Landforms of the Louisiana Coastal Plain*, Louisiana State University, 2019.

105 *eating away at the sedimentary land*: Tyler Kelley, *Holding Back the River: The Struggle Against Nature on America's Waterways* (New York: Simon & Schuster, 2021). See also John Barry, *Rising Tide* (New York: Simon & Schuster, 1997).

105 *watery maze is called "the bayou"*: William Conner and John Day, *The Ecology of Barataria Basin, Louisiana: An Estuarine Profile*, US Department of the Interior, Fish and Wildlife Service, July 1987.

106 *dominant mainland tribes such as the Chitimacha*: "Final Environmental Assessment: Pointe-au-Chien Cultural Heritage Protection Reef in Coastal Louisiana," National Oceanic and Atmospheric Administration Office for Coastal Management, February 2019.

106 *a permanent movement south*: "Native Americans: Houma Indians," Terrebonne Parish GenWeb, accessed November 2021, http://lagenweb.org/terrebonne/resources/indi ans.html.

106 *find an undisturbed refuge*: "History of Choctaws in Louisiana," historical article produced for choctawnation.com. More information can be found in Fred Kniffen et al., *The Historic Indian Tribes of Louisiana: From 1542 to the Present* (Baton Rouge: LSU Press, 1994).

106 *colonial administrator Jean-Baptiste le Moyne de Bienville*: "Tribal History," Sovereign Nation of the Chitimacha, accessed November 2021, http://www.chitimacha.gov /history-culture/tribal-history. See also Kniffen et al., *The Historic Indian Tribes of Louisiana*.

106 *what is now Terrebonne Parish*: J. Daniel D'Oney, "The Houma Nation: A Historiographical Overview," *Louisiana History: The Journal of the Louisiana Historical Association* 47, no. 1 (Winter 2006): 63–90.

106 *more than two-thirds water*: "Terrebonne Parish," landing page on LA Safe website. LA Safe is the Louisiana state government project to facilitate migration away from the coast and adaptation to climate change.

106 *had the land to themselves*: "Louisiana: European Explorations and the Louisiana Purchase—A Special Presentation from the Geography and Map Division of the Library of Congress," Library of Congress, accessed October 2021, https://www .loc.gov/collections/louisiana-european-explorations-and-the-louisiana-purchase /about-this-collection/.

107 *came to be known as "Cajuns"*: Christopher Hodson, *The Acadian Diaspora: An Eighteenth-Century History*, reprint ed. (New York: Oxford University Press, 2017).

107 *an Indigenous woman*: "Alexandre Verdun," page within Pointe-aux-Chenes family tree compiled by Michele Dardar on Geneanet. I am grateful to Michele for his meticulous work compiling the lineages of Pointe-aux-Chenes and Isle de Jean Charles.

See also "Summary Under the Criteria and Evidence for Amended Proposed Finding Against Federal Acknowledgment of the Pointe-au-Chien Indian Tribe," Bureau of Indian Affairs Office of Federal Acknowledgment, May 22, 2008, 8–9, 47–50.

107 *around Bayou Pointe-au-Chien*: Roberta Estes, "Houmas Indians of Terrebonne Parish, Louisiana," Native Heritage Project, October 27, 2012.

107 *interracial marriages during that period*: Bureau of Indian Affairs Office of Federal Acknowledgment 2008, 47–50. See also "Summary Under the Criteria and Evidence for Proposed Finding Against Federal Acknowledgment of the United Houma Nation," US Bureau of Indian Affairs Office of Federal Acknowledgment, December 13, 1994, 65–71.

107 *Indigenous ritual and Catholic sacrament*: Laura Browning, *Faith, Families, & Friends: 150 Years of Sacred Heart of Jesus Parish and Montegut, Louisiana* (Morrisville, NC: Lulu Press, 2016), ch. 1, "Prior to 1860."

107 *just four common surnames*: "Summary Under the Criteria and Evidence for Proposed Finding Against Federal Acknowledgment of the United Houma Nation," 35.

108 *cypress rafts and dugout canoes*: Interviews with Christine Verdin, Albert Naquin, and Jake Billiot, August 2021.

108 *protect their villages from storms*: Photographs provided by Christine Verdin and the Pointe-au-Chien Indian Tribe, also from interview with Christine Verdin, August 2021.

108 *washed up over the islands*: Interview with Jake Billiot, August 2021.

108 *weave baskets out of palmetto fronds*: Interview with Roch Naquin, July 2021. See also Nathalie Dajko, *French on Shifting Ground: Coastal and Cultural Erosion in South Louisiana* (Jackson: University Press of Mississippi, 2020).

108 *"carrots, corn, and rice"*: Julie Maldonado, *Facing the Rising Tide: Co-Occurring Disasters, Displacement, and Adaptation in Coastal Louisiana's Tribal Communities* (Washington, DC: American University, 2014), 76.

109 *in today's money*: Interviews with Roch Naquin and Theresa Dardar, June and July 2021.

109 *same was true for oysters*: Interviews with Jake Billiot and Chuckie Verdin, August 2021; see also Maldonado, *Facing the Rising Tide*, 33–37, 87.

109 *large sugarcane plantations*: Interviews with Roch Naquin, Chuckie Verdin, and Don Ellender, August 2021.

110 *pushing east toward New Orleans*: Tommy Stringer, "Joseph S. Cullinan: Pioneer in Texas Oil," *East Texas Historical Journal* 19, no. 2 (October 1981): 43–59.

110 *refine it into gasoline*: G. D. Harris, "Oil and Gas in Louisiana with a Brief Summary of Their Occurrence in Adjacent States," United States Geological Survey, 1910.

110 *the company's trapping areas*: Final decree in *Delaware Louisiana Fur Trapping Company v. Bizani et al.*, accessed in Lafourche Parish District Court, Thibodeaux, Louisiana.

Record contained in evidence for Civil Suit 72464, *Louisiana Land & Exploration v. Verdin* (1995).

110 *drill derricks and pipelines*: James Sell and Tom McGuire, "History of the Offshore Oil and Gas Industry in Southern Louisiana—Volume IV: Terrebonne Parish," US Department of the Interior OCS Study, September 2008.

111 *see the derricks from their yards*: Maldonado, *Facing the Rising Tide*, 117–18.

111 *keep off the company's land*: Permanent injunction in *The Louisiana Land and Exploration Company v. Gazo Cheramie et al.*, federal lawsuit filed in November 1933, accessed in Lafourche Parish District Court, Thibodeaux, Louisiana. Record contained in evidence for Civil Suit 72464, *Louisiana Land & Exploration v. Verdin* (1995). Context provided in interview with Joel Waltzer, attorney representing the Pointe-au-Chien Indian Tribe.

111 *the Mississippi River*: These two waterways were the Houma Navigational Canal, constructed by the Army Corps of Engineers, and the Intracoastal Waterway, part of a system of shipping channels that runs along the Gulf of Mexico from Texas to Florida.

112 *slide down into the water*: Julie Bernier, "Induced Subsidence Related to Hydrocarbon Production—Subsidence and Wetland Loss Related to Fluid Energy Production, Gulf Coast Basin," Report for the US Geological Survey by the St. Petersburg Coastal and Marine Science Center, June 7, 2020.

112 *stop the water from rising*: Tyler Kelley, *Holding Back the River*.

112 *screw into a wall*: Barry Yeoman, "'The Land Is Washing Back to the Sea.'" onEarth, November 6, 2010, part of Yeoman's extraordinary ten-part series *Losing Louisiana*. Historical maps of erosion in Terrebonne Parish also provided by Christine Verdin from archives of the Pointe-au-Chien Indian Tribe.

112 *Bayou de la Valle, and more*: Dayna Bowker Lee, "Louisiana Indians in the 21st Century," Folklife in Louisiana, accessed October 2021, https://www.louisianafolklife.org/LT/Articles_Essays/nativeams.html. See also Dajko, *French on Shifting Ground*.

112 *larger than the state of Delaware*: "Louisiana's Changing Coastal Wetlands," news release from the US Geological Survey, July 12, 2017.

114 *Marie the* femme sauvage: Pointe-aux-Chenes family tree compiled by Michele Dardar on Geneanet, also interviews with Chuckie Verdin and Christine Verdin, August 2021.

114 *and so on*: Interviews with Alton Verdin, Jake Billiot, and Jacco Billiot, August 2021.

114 *along with double-wide trailers*: Photographs and oral histories provided by Christine Verdin and Missy Billiot for the Pointe-au-Chien Indian Tribe.

115 *hooves lodged in soaking mud*: Interviews with Roch Naquin, Jake Billiot, and Theresa Dardar, July and August 2021.

115 *poisoned by the salt water*: Interview with Theresa Dardar, July 2021. See also Barry Yeoman, "Reclaiming Native Ground: Can Louisiana's Tribes Restore Their Traditional Diets as Waters Rise?," *The Lens*, February 9, 2017.

115 *before he got another day off*: Sell and McGuire, "History of the Offshore Oil and Gas Industry in Southern Louisiana—Volume IV: Terrebonne Parish," 18.

115 *farther along to Texas*: Interviews with Christine Verdin, August and September 2021.

116 *middlemen who came down from New Orleans*: Interview with Jake Billiot, August 2021.

116 *exceeded a few thousand*: Elizabeth LaFleur, Diane Yeates, and Angelina Aysen, "Estimating the Economic Impact of the Wild Shrimp, Penaeus sp., Fishery: A Study of Terrebonne Parish, Louisiana," *Marine Fisheries Review* 67, no. 1 (Winter 2005): 28–42.

116 *to make ends meet*: Interviews with Jake Billiot and Chuckie Verdin, August and September 2021.

116 *pounds of fresh catch*: Interview with Rosie Dardar, resident of Isle de Jean Charles and a longtime worker at a shrimp peeling plant, April 2021.

116 *toward the marsh for their first catch*: Interviews with Roch Naquin and Timmy Kerner, mayor of the town of Jean Lafitte in Jefferson Parish. See also Betsy Gordon, "Decorating for the Shrimp Fleet Blessing: Chauvin, Louisiana," Folklife in Louisiana, accessed October 2021, https://www.louisianafolklife.org/LT/Articles_Essays/main_misc_shrimp_fleet_dec.html. I watched the Pointe-aux-Chenes boat blessing in April of 2021.

117 *only for whites*: Interviews with Chuckie Verdin, Christine Verdin, and Don Ellender, August 2021.

117 *effects of erosion*: Laise Ledet, *They Came, They Stayed: Origins of Pointe-aux-Chenes and Ile à Jean Charles: A Genealogical Study, 1575-1982*. Self-published, 1982. The resonance is not an intentional reference to the land's immunity to erosion but rather a historical coincidence.

118 *a narrow island in the marsh*: Interviews with Alton Verdin, Mary Verdin, Sheri Neil, ShanaRae Dardar, and numerous others, April through September 2021.

118 *after class let out*: Interviews with Chuckie Verdin, July through September 2021.

118 *in the hospital with burn wounds*: Interview with Christine Verdin, August 2021.

119 *out of the deep waters*: Interviews with Chuckie Verdin, July through September 2021.

119 *perched on stilts*: Kimberly Solet, *Thirty Years of Change: How Subdivisions on Stilts have Altered a Southeast Louisiana Parish's Coast, Landscape and People*, University of New Orleans master's thesis, 2006.

120   *spiraling out into the marsh*: Interview with Chuckie Verdin, August 2021.

121   *something like a community elder*: Interview with Christine Verdin; see also Office of Federal Acknowledgment 2008, "Summary Under the Criteria and Evidence for Amended Proposed Finding Against Federal Acknowledgment of the Pointe-au-Chien Indian Tribe," 28–29.

121   *his small herd of cattle*: Interviews with Alton Verdin and Jake Billiot; see also "Summary Under the Criteria and Evidence for Amended Proposed Finding Against Federal Acknowledgment of the Pointe-au-Chien Indian Tribe," 44.

121   *oil pipeline and a gas pipeline*: LL&E holdings map filed in evidence in docket for *LL&E v. Verdin* (1995), records accessed in Lafourche Parish Courthouse, Thibodeaux, Louisiana.

121   *he had already rebuilt it*: LL&E daily and monthly patrol reports filed in evidence in docket for *LL&E v. Verdin* (1995).

122   *walked out scot-free*: Interviews with Alton Verdin, Jake Billiot, and Jacco Billiot, August 2021. Gary passed away in 2019.

123   *sabotage against the company*: Final order in *LL&E v. Verdin* (1995). Context provided by Joel Waltzer, attorney for the Pointe-au-Chien Indian Tribe.

123   *would soon disappear altogether*: Interviews with Chuckie Verdin, July through September 2021.

124   *swampland homes erode away*: "For Gullah Geechee People on the SC Coast, Climate Change Is Already a Threat," WFAE, October 28, 2021.

124   *battered by storm surge*: Teresa Tomassoni, "The Rising Pacific Forces a Native Village to Move. Who Will Pay?" *Bloomberg*, November 5, 2021.

124   *forced to move by rising seas*: Chris Mooncy, "The Remote Alaskan Village That Needs to Be Relocated Due to Climate Change," *Washington Post*, February 24, 2015.

124   *and thawing permafrost*: Julia Ilhardt, " 'It Was Sad Having to Leave': Climate Crisis Splits Alaskan Town in Half," *The Guardian*, June 8, 2021.

124   *that would be enough*: "Summary Under the Criteria and Evidence for Amended Proposed Finding Against Federal Acknowledgment of the Pointe-au-Chien Indian Tribe," 43–45.

125   *a decade of back-and-forth*: "Summary Under the Criteria and Evidence for Proposed Finding Against Federal Acknowledgment of the United Houma Nation," US Bureau of Indian Affairs Office of Federal Acknowledgment, December 13, 1994.

125   *the council of elders*: Interviews with Chuckie Verdin and Christine Verdin, August 2021.

126   *"required for Federal acknowledgment"*: "Summary Under the Criteria and Evi-

dence for Amended Proposed Finding Against Federal Acknowledgment of the Pointe-au-Chien Indian Tribe," 1.

126 *for cut-rate prices*: Peter Debaere, *Small Fish—Big Issues the Effect of Trade Policy on the Global Shrimp Market*, Darden Business School, University of Virginia, February 2008. The common view that Asian countries "dumped" their shrimp in the United States represents an oversimplification of global trade dynamics, but nevertheless the volume of imports did increase significantly in the early 2000s.

126 *two dollars a pound or less*: Chris Kirkham, "Shrimp Season Opens Today, but Crush of Imports Has Profoundly Altered LA Industry," *Times-Picayune*, May 10, 2009. See also Brian Marks, "The Political Economy of Household Commodity Production in the Louisiana Shrimp Fishery," *Journal of Agrarian Change* 12, nos. 2–3 (April/July 2012): 227–51. Also from interviews with Chuckie Verdin and Jake Billiot, August 2021.

126 *earn back the cost of fuel*: Interviews with Chuckie Verdin, July through September 2021.

127 *no longer sustain him*: Interviews with Chuckie, Christine, Charlie, and Angele Verdin, July through September 2021.

127 *giving up his home*: Interviews with Charlie and Angele Verdin, July and August 2021.

128 *pushed more people to leave the bayou*: Interviews with Robert Verdin, Chuckie Verdin, and Christine Verdin, August 2021.

128 *look for other work*: Campbell Robertson, "Determined to Make a Living Before Oil Arrives," *New York Times*, May 11, 2010.

128 *plants that held the marsh in place*: Alessandra Potenza, "The 2010 Deepwater Horizon Oil Spill Caused Widespread Land Erosion in Louisiana," *The Verge*, November 23, 2016.

128 *far from what it had once been*: Interviews with Chuckie Verdin and Jake Billiot, August 2021.

128 *in special nets*: Patrick Oppmann, "Oil Spill Threatens Native American Land," CNN, June 1, 2010. Also interviews with Chuckie Verdin and Jake Billiot, August 2021.

128 *spill caused to the bayou wetlands*: Laurel Brubaker Calkins and Allen Johnson Jr., "American Indian Tribe Sues BP for Oil Spill Damages," *Bloomberg Businessweek*, April 18, 2011.

130 *you were out of luck*: Interview with Charlie Verdin, July 2021.

130 *progress of erosion on the bayou*: Tour of tribal building provided by Christine Verdin and Theresa Dardar, August 2021.

130 *singing lessons from tribal elders*: Kezia Setyawan, "Pointe-au-Chien Children Gain Knowledge and Pride at Culture Camp," *Houma Courier*, July 24, 2021.

130 *draw people back to the bayou*: Interview with Christine Verdin, August 2021.

131 *effective that summer*: Interviews with Sheri Neil and ShanaRae Dardar at Sheri's Snack Shack, April 2021.

132 *lawsuit against the parish*: Kezia Setyawan, "Pointe-aux-Chenes Parents File Federal Lawsuit over Elementary School Closure," *Houma Courier*, June 18, 2021.

132 *holding up handwritten signs*: Kezia Setyawan, "Pointe-aux-Chenes Parents, Residents and Students Protest Proposed School Closure," *Houma Courier*, April 1, 2021.

132 *empty desks in every classroom*: Interviews with Natalie Bergeron and Mary Verdin, April and August 2021.

132 *their unique traditions to the next*: Interviews with Alton Verdin, Mary Verdin, and ShanaRae Dardar, April through August 2021.

133 *aftermath of Hurricane Katrina*: Kezia Setyawan, "Officials Tour Parts of the Local Morganza-to-the-Gulf Hurricane Protection System," *Houma Courier*, August 22, 2021.

133 *safe for decades to come*: Interviews with Christine Verdin and Theresa Dardar, August 2021.

133 *new tract of land farther inland*: Kezia Setyawan, "Isle de Jean Charles Residents View Future Homes for the First Time," *Houma Courier*, April 7, 2021. There has been extensive coverage of the Isle de Jean Charles relocation in many major media outlets and in Elizabeth Rush's excellent book *Rising: Dispatches from the New American Shore* (Minneapolis: Milkweed, 2018). I am grateful to Albert Naquin, Roch Naquin, and Rosie Dardar of Isle de Jean Charles for sharing their perspectives on the relocation.

134 *to shut the school down*: Video of meeting provided by Andy Metzger.

134 *on both his mother's and his father's side*: Interview with Alton and Mary Verdin, August 2021.

136 *double-wide trailers*: Videos provided by Jacco Billiot.

136 *white wood all down the block*: Author's own observations from a visit to Pointe-aux-Chenes on August 30, 2021, the day after Hurricane Ida made landfall. I am grateful to Kezia Setyawan for accompanying me on the drive back into Terrebonne Parish in the aftermath of the storm.

137 *elsewhere in Terrebonne Parish*: Interview with Alton Verdin, August 2021.

137 *as a French language magnet school*: Kezia Setyawan, "French immersion school in Pointe-aux-Chenes to open in 2023 after unanimous bill approval," WWNO, June 14, 2022.

CHAPTER 5: FRANKENSTEIN CITY

141 *embargo on exports to the US*: "The Yom Kippur War: 40 Years of Survival," Richard Nixon Foundation, October 11, 2016.

141 *prices stateside to skyrocket*: "Oil Embargo, 1973–1974," US State Department Office of the Historian, accessed October 2021, https://history.state.gov/milestones/1969 -1976/oil-embargo.

141 *supply gap created by the embargo*: Michael Corbett, "Oil Shock of 1973–74," Federal Reserve History, November 22, 2013.

141 *Christmas light displays*: Andrew Malcolm, "Fuel Crisis Dims Holiday Lights," *New York Times*, November 25, 1973.

141 *outpost into a boomtown*: David Pitman, "OPEC Plus 40—How Houston Changed, Then and Now," Houston Public Media, October 21, 2013.

142 *the morning rush hour*: Abby Livingston, "'All of the Party Was Over': How the Last Oil Bust Changed Texas," *Texas Tribune*, May 18, 2020.

142 *stumbled out in every direction*: William Stevens, "Houston, Fastest Growing Big City, Showing Signs of Having Hit Prime," *New York Times*, December 16, 1981.

142 *neighborhoods wrestled for space*: Jay Jordan, "Maps Show How Houston Has Grown Since 1836," *Houston Chronicle*, November 13, 2019.

142 *every inch of undeveloped prairie*: Neena Satija, Kiah Collier, and Al Shaw, "Boom-town, Flood Town," *Texas Tribune*/ProPublica, December 13, 2016.

142 *the northwest edge of the city*: "Consent to Subdivision Plat for Woodland Trails Section 1," filed March 31, 1975. Records available courtesy of the Harris County Clerk's Office.

143 *"Architectural Control Committee"*: "Restrictions, Covenants, and Conditions for Woodland Trails Section One," filed October 5, 1967. Records available courtesy of the Harris County Clerk's Office. There was another subdivision right next door, this one called Woodland Oaks and owned by a different developer, but the two neighborhoods were so similar that most residents thought of them as a single unit.

143 *seldom any outsiders passing through*: Interviews with Cindy Mazzola, Lois Fisher, Reginald Smith, and Ron McClain, August 2021.

143 *mounds of concrete*: "White Oak Bayou: History of a Houston Waterway," White Oak Bayou Association, accessed October 2021, http://whiteoakbayou.org/uploads/3/4 /9/1/34911613/historysection.pdf.

143 *real restrictions on permitting*: Michael Hagerty, "Exploring the Power of—and Potential Problems with—Municipal Utility Districts," Houston Public Media, August 30, 2016.

143 *going through such boom times*: Interview with John Blount, chief engineer of Harris County, June 2021. I am grateful to John for his perspective on the history of Houston's development and the Harris County buyout program.

144 *sloshed down streets and sidewalks*: Interviews with Cindy Mazzola, Terry Rayford Rollins, and Esther Del Toro, August 2021.

144 *might otherwise fizzle out*: Carrigan Chauvin, "Remembering Tropical Storm Allison," CW39 Houston, June 7, 2021.

144 *freeways and tollways*: "Service Assessment: Tropical Storm Allison Heavy Rains and Floods, Texas and Louisiana, June 2001," US Department of Commerce, National Oceanic and Atmospheric Administration, June 2001.

145 *to buy insurance*: National Research Council, *Affordability of National Flood Insurance Program Premiums: Report 1* (Washington, DC: National Academies Press, 2015).

146 *chance of happening in each year*: Rebecca Elliott, *Underwater: Loss, Flood Insurance, and the Moral Economy of Climate Change in the United States* (New York: Columbia University Press, 2021), ch. 1. Also from interviews with Rebecca Elliott and A. R. Siders, March through September 2021.

146 *and toward higher ground*: Interview with John Blount, June 2021.

146 *any other such program in the country*: Jake Bittle, "On the Waterfronts: Flood Buyouts and the Economics of Climate Catastrophe," *The Baffler* 49 (January 2020): 124–33.

147 *the now-affordable subdivisions*: Interviews with Cindy Mazzola and Galia Vargas, August 2021.

147 *get in the way of their stability*: Interview with Cindy Mazzola, August 2021.

148 *surrounding postal code*: Robert Benincasa, "Search the Thousands of Disaster Buyouts FEMA Didn't Want You to See," NPR, March 5, 2019.

149 *where they ended up*: Kevin Loughran and James R. Elliott, "Residential Buyouts as Environmental Mobility: Examining Where Homeowners Move to Illuminate Social Inequities in Climate Adaptation," *Population and Environment* 41, no. 1 (September 2019); and James Elliott et al., "Divergent Residential Pathways from Flood-Prone Areas: How Neighborhood Inequalities Are Shaping Urban Climate Adaptation," forthcoming in *Social Problems*. Thank you to Jim Elliott for providing very helpful context on this invaluable research.

150 *played music at odd hours*: For obvious reasons, this person did not wish me to use their name.

150 *gang activity in his high school*: Interview with Cindy Mazzola, August 2021.

150 *less leverage to decide where they would move*: Interviews with Galia Vargas, Gloria Riley, Laura Mendez, Rosa Hernandez, and Lois Fisher, April through September 2020 and June through August 2021.

150 *raise their two young daughters*: Interview with Audelina Gomez, August 2021. I am

deeply grateful to my research assistant Maddie Parrish for conducting this interview in Spanish.

151 *they could omit flooding*: Alex Harris, "Under Texas Law, Homeowners Warned of Potential Flooding—Not in Florida," *Miami Herald*, October 26, 2019. Also from interview with James Wade, manager of the Harris County Flood Control District's home buyout program.

154 *incentivized them to stay*: "Report: Strategic Property Buyouts to Enhance Flood Resilience," Nature Conservancy, March 2019. Also from interviews with James Wade, Harris County Flood Control District buyout program manager, November 2019 and May 2020.

154 *advertising its large "side yard"*: Interview with James Wade, May 2020.

154 *often took three to five years to execute*: Anna Weber and Rob Moore, "Going Under: Long Wait Times for Post-Flood Buyouts Leave Homeowners Underwater," Natural Resources Defense Council, September 12, 2019.

154 *left their successors to deal with the consequences*: Parcel records from the Harris County Appraisal District for the remaining homes in Woodland Trails West show that the large majority of still-standing homes were sold in the years immediately after Tropical Storm Allison, and sometimes sold again several times in the years that followed. This allows us to conclude that most residents who still live in the neighborhood were not present for the 1999 and 2001 floods.

154 *short-term tenants who came and left*: Interview with James Wade.

155 *spacious and empty*: Interviews with several other residents of Woodland Trails West. During a visit to the neighborhood in August of 2021 I encountered a few groups of Frisbee golf players.

156 *rivaled that of Chicago*: Paul Debenedetto, "Harris County's Population Grew More Than 15% in the Last Decade," Houston Public Media, August 13, 2021.

156 *crossed county lines, and kept going*: Jay Jordan, "Maps Show How Houston Has Grown Since 1836."

156 *a hundred years' worth of good data*: Robert E. Hinshaw, *Living with Nature's Extremes: The Life of Gilbert Fowler White* (Boulder, CO: Johnson Books, 2006). This is a loving biography of Gilbert White, the father of modern flood insurance and land use policy. Also from interview with Rebecca Elliott. See also "Reducing Flood Losses: Is the 1% Chance (100-Year) Flood Standard Sufficient?" Report prepared for the 2004 Assembly of the Gilbert F. White National Flood Policy Forum, September 21–22, 2004.

156 *thousand-year flood event*: Jason Samenow, "Harvey Is a 1,000-Year Flood Event Unprecedented in Scale," *Washington Post*, August 31, 2017.

157 *been trying to buy out*: Lisa Song, Al Shaw, and Neena Satija, "After Harvey, Buyouts Won't Be the Answer for Frequent Flood Victims in Texas," *Texas Tribune*, November 2, 2017.

157 *in the years before Harvey*: Mike Morris et al., "Meyerland Floods Again; Residents Wonder, 'What Now?'" *Houston Chronicle*, September 2, 2017.

157 *elevated on wooden stilts*: Dolores Mendoza and Gabrielle Luebano, "Opinion: The County Can Have Our Homes of 6 Generations—but Not Our Dignity," *Houston Chronicle*, August 29, 2020.

157 *filled with filthy water*: Dylan McGuinness, "Council OKs $11.5 Million Buyout of Apartment Complex to Boost Flood Protection in Westbury," *Houston Chronicle*, November 18, 2020.

158 *"reasonably expected or provided against"*: Marco Ornelas and Rebecca Andrews, "Court of Federal Claims Rejects Takings Claims Related to Hurricane Harvey Downstream Flooding Cases," Argent Communications Group, March 28, 2020.

158 *retreat from the bayous*: Historical screenshots of Wimbledon Champions West area, viewable on Google Earth Pro.

158 *happening each year*: The regulatory floodplain boundaries for the area can be viewed at FEMA's Flood Map Service Center, located at msc.fema.gov.

159 *hundred-year floodplain threshold*: Alexandra Tempus, "Presto Chango: How Flood Map Revisions Allow Building in Risky Areas," FairWarning, November 12, 2020.

159 *filled with brown water*: John Schwartz, James Glanz, and Andrew Lehren, "Builders Said Their Homes Were out of a Flood Zone. Then Harvey Came," *New York Times*, December 2, 2017.

159 *close to the city center*: Interviews with James Wade and John Blount, May 2020 and June 2021.

159 *rainfall on dry concrete*: Christina Rosales, "Houston Knew Neighborhoods of Color Were Inadequately Protected from Even Modest Storm Events," Texas Housers, August 31, 2017. See also "Sunnyside and South Park Comprehensive Needs Assessment Data Report," produced by the Baker Institute at Rice University, November 2019.

160 *for bed rest and induced labor*: Interview with Becca and Sergio Fuentes, August 2021.

162 *spraying along the length of the street*: This chapter's account of the flood in Bear Creek Village comes from interviews with Becca and Sergio Fuentes, Carl Gabbard, Linda Kay Beaver, Gail O'Neil, and Medy Onia, and other residents, all conducted in August and September 2021.

164 *was ever going to prosper*: Katherine Arcement, "Houston's Besieged Dams Were Built

70 Years Ago After This 1935 Flood Devastated the City," *Washington Post*, August 31, 2017.

164 *moat for the rest of the city*: Michael Bloom, "The History of Addicks and Barker Reservoirs," *Riparian Houston*, September 3, 2017.

165 *its engineers wrote in 1986*: Sara Frosch and Dan Randazzo, "Homeowners Fault Government for Hurricane Harvey Damage," *Wall Street Journal*, May 6, 2019.

165 *did not have that long to wait*: There ended up being two separate lawsuits against the Army Corps of Engineers over the reservoir flooding, one from the "upstream plaintiffs," who lived west of the overtopped reservoir, and one from the "downstream plaintiffs," whose homes had been flooded by releases from the reservoir. The courts found in favor of the upstream plaintiffs, including residents of Bear Creek Village, but found against the downstream plaintiffs. At the time of this writing, damages in the former case have yet to be paid out.

166 *an endless procession of contractors*: Interview with Becca and Sergio Fuentes, August 2021.

168 *rent them out to newcomers*: Prashant Gopal, "Distressed Investors Are Already Buying Houston Homes for 40 Cents on the Dollar," *Bloomberg Businessweek*, October 12, 2017.

168 *largest owner of such properties*: Francesca Mari, "A $60 Billion Housing Grab by Wall Street," *New York Times Magazine*, March 4, 2020.

168 *institutional investors and hedge funds*: David Hunn and Matt Dempsey, "In Houston's Flooded Neighborhoods, Real Estate Investors See an Opportunity," *Houston Chronicle*, May 10, 2018.

168 *"We will kill him"*: Adam Lewis, "Private Equity Should Become Less Private to Address Onslaught of Criticism," PitchBook, December 10, 2019.

169 *a little behind on rent*: Todd Frankel and Dan Keating, "Eviction Filings and Code Complaints: What Happened When a Private Equity Firm Became One City's Biggest Homeowner," *Washington Post*, December 25, 2018.

169 *sell many of those properties to other investors*: Harris County Appraisal District records show that Cerberus owned more than 1,100 properties in Harris County as of the end of the 2017 tax year, but that number shrank in subsequent appraisal years as the firm sold off bundles of its homes.

169 *Juan had kept paying*: Interview with Juan Torres, August 2021.

170 *"some folks' intention to move out of the area"*: Interview with Dustin Gaspari. Harris County Appraisal District records show transactions by Gaspari's company Agave were concentrated in Bear Creek and surrounding neighborhoods.

170   *home transactions died down*: Interviews with several Bear Creek residents including Medy Onia, Gail O'Neal, and Juan Torres, August 2021.

170   *owned by Cerberus Capital Management*: Interview with Jacob Lee, August 2021.

172   *for buyouts and other projects*: Adam Bennett, "Harris County $1.4 Billion Short on Bond Flood Control Projects," KHOU, May 19, 2021.

172   *fix the Addicks and Barker Reservoirs*: Travis Bubenik, "Army Corps Plans to Study Improvements to Addicks and Barker Reservoirs," Houston Public Media, February 22, 2018.

172   *right combination of excitement and privacy*: Interview with Medy Onia, August 2021.

174   *storm drains of ample size*: Author's own observations from visit to Grand Oaks in August of 2021. See also FEMA Flood Map Service Center, msc.fema.gov.

174   *eclipsing Allison's record once again*: Michelle Iracheta, "The Highest Rainfall Totals from Tropical Storm Imelda Across Southeast Texas," *Houston Chronicle*, September 23, 2019.

CHAPTER 6: WHY SHOULD THIS A DESERT BE?

179   *for some well-deserved rest*: Interview with Karen Felkins, July 2021.

180   *the last twelve hundred years*: Alejandra Borunda, "'Megadrought' Persists in Western U.S., as Another Extremely Dry Year Develops," *National Geographic*, May 7, 2021.

180   *They had never seen cows behave that way before*: Interview with Karen Felkins, July 2021.

181   *feed their penned-up herds*: Interview with Ben Dickman, dairyman in Pinal County.

181   *many ranchers were walking away with pennies*: Jaweed Kaleem, "Starving Cows. Fallow Farms. The Arizona Drought Is Among the Worst in the Country," *Los Angeles Times*, August 3, 2021.

182   *The younger generation wasn't interested in taking over*: Interview with Wyatt Ferreira, another rodeo rancher in Pinal County, July 2021.

182   *which would lead to fewer wildfires*: Lili Pike, "What the Megadrought in the West Means for Wildfire Season," *Vox*, April 15, 2021.

182   *"heat dome" of the summer of 2021 much less likely*: Umair Irfan, "The Surprisingly Subtle Recipe Making Heat Waves Worse," *Vox*, June 23, 2021.

183   *boiling salmon alive in their streams*: Emily Chung, "Salmon Are Getting Cooked by Climate Change. Here's How They Could Be Saved," CBC News, July 23, 2021.

183   *to account for declining river flows*: Chris Baker, "Sacramento Asking People to Reduce Water Use by 15%, Doubles Fines for Water Waste," ABC10, August 24, 2021.

183   *planting the wheat in the first place*: "Drought Having Serious Impact on Idaho's Wheat Crop," Idaho News 6, October 6, 2021.

183 *"prayer and fasting for drought relief"*: Kevin Miller, "Canyon County Officials 'Fast and Pray for Rain Sunday,'" KIDO Radio, October 9, 2021.

184 *in the scalding hundred-degree weather*: "Sonoran Desert Network Ecosystems," US National Park Service, accessed November 2021, https://www.nps.gov/im/sodn/eco systems.htm.

184 *centered around the domestication of corn*: William L. Merrill et al., "The Diffusion of Maize to the Southwestern United States and Its Impact," *Proceedings of the National Academy of Sciences* 106, no. 50 (December 2009): 21019–26.

185 *where it seeped across the fields*: Kim Whitley and Jeri Ledbetter, "Hohokam Canal System," Arizona Heritage Waters, a project of Northern Arizona University, accessed November 2021, http://www.azheritagewaters.nau.edu/loc_hohokam.html.

185 *"for irrigating their fields"*: Father Pedro Mendez, quoted in Charles Polzer, *The Jesuit Missions of Northern Mexico* (New York: Garland, 1991).

185 *a more marketable name—Phoenix*: Patricia Gober, *Metropolitan Phoenix: Place Making and Community Building in the Desert* (Philadelphia: University of Pennsylvania Press, 2013).

185 *"one of the handsomest in the West"*: "Bird's Eye View of Phoenix, Maricopa Co., Arizona," map dated circa 1885, Library of Congress, https://www.loc.gov/resource /g4334p.pm000110/.

185 *"reclaim" the desert land for farming*: "The Bureau of Reclamation: A Very Brief History," US Bureau of Reclamation, accessed November 2021, https://www.usbr.gov /history/borhist.html.

186 *saw fit to let white settlers tap it, too*: *Draft Environmental Assessment San Carlos Irrigation Project Facilities Phase 2 Rehabilitation, Reaches 1–3*, US Bureau of Reclamation, May 2017.

186 *network of canals that transected the county*: Interview with Nancy Caywood, descendant of Lewis Storey, September 2021. Contemporaneous photographs and records provided by Nancy Caywood.

186 *about six hundred thousand gallons*: An acre-foot of water is the amount of water it takes to cover an acre of land in a foot of water.

186 *ideal for agriculture*: Julie Murphree, "'Why in God's Name Are We Growing Cotton in the Desert?,'" Arizona Farm Bureau, May 18, 2016.

187 *you could smell it from half a mile away*: Interview with Nancy Caywood, also attested by Kathy Wuertz, September 2021.

187 *thousands upon thousands of cotton bales*: Interviews with Nancy Caywood and Paco Ollerton, September 2021.

187 *a gold rush of so-called groundwater*: *Layperson's Guide to Arizona Water*, Water Education Foundation and University of Arizona Water Resources Research Center, 2007.

187 *of its agricultural water supply*: "Water Uses in 1960s Arizona," US Bureau of Reclamation, accessed November 2021, https://www.usbr.gov/lc/phoenix/AZ100/1960/water_uses_1960.html.

188 *depleting the aquifers faster than they could regenerate*: Robert G. Dunbar, "The Arizona Groundwater Controversy at Mid-Century," *Arizona and the West* 19, no. 1 (Spring 1977): 5–24.

188 *what amounted to probation*: Brian McGreal and Sussana Eden, "Arizona Groundwater Management: Past, Present and Future," *Arroyo 2021*, University of Arizona Water Resources Research Center, 2021. See also Jon Kyl, "Arizona's New Groundwater Statute: 1980 Groundwater Management Act," *Water Resources Allocation: Laws and Emerging Issues*, June 8, 1981. Kyl later went on to serve as US senator for Arizona for almost two decades.

188 *the other cotton producers in the county*: Interviews with Nancy Caywood and Paco Ollerton.

188 *who could use its water and for how long*: Marc Reisner, *Cadillac Desert: The American West and Its Disappearing Water*, rev. ed. (New York: Penguin Books, 1993).

188 *"lower basin" states like Arizona*: David Owen, *Where the Water Goes: Life and Death Along the Colorado River* (New York: Riverhead, 2017).

188 *most junior of any state*: "Sharing Colorado River Water History, Public Policy and the Colorado River Compact," *Arroyo 1997*, University of Arizona Water Resources Research Center, 1997.

188 *hundreds of miles away from Phoenix*: Thomas McCann, "Central Arizona Project: A Brief History," presentation prepared for Central Arizona Project, April 26, 2013. See also notes 35 and 36.

189 *vast expanses of farmland in Pinal County*: "System Map," Central Arizona Project, accessed November 2021, https://www.cap-az.com/water/cap-system/water-operations/system-map/.

189 *the farmers of Pinal County had all the water they needed*: Interviews with Brady Udall at the University of Utah and Sarah Porter at Arizona State University, two academics who have studied water rights in Arizona for decades, January and March 2021. See also Joe Gelt, "Long-Awaited CAP Delivers Troubled Waters to State," *Arroyo* 6, no. 3, University of Arizona Water Resources Research Center, fall 1992.

189–90 *replicate this robust enterprise on his own land*: Interview with Cassy England, Don's granddaughter, September 2021.

190 *concrete ditches from field to field*: "About Us: Organization, Operation, and Description

of CAIDD and Electrical District No. 4," Central Arizona Irrigation and Drainage District, accessed November 2021, https://www.ed4.biz/view/43/.

190 *the stalks of cotton sprang up*: Interviews with Cassy England, Paco Ollerton, Kathy Wuertz, and Nancy Caywood, September 2021.

191 *that farming had a future in Pinal County*: Interviews with Paco Ollerton and Cassy England, September 2021.

191 *caprices of the cotton commodity market*: Katie Campbell, "With Cotton Marginal, Some Farmers Choose Retirement," *Tri-Valley Dispatch*, October 12, 2016.

193 *she would not be receiving any water after all*: Letter from San Carlos Irrigation and Drainage District provided courtesy of Nancy Caywood.

193 *before high cotton season arrived*: Interview with Nancy Caywood, September 2021.

193 *dwindled to around fifty acre-feet*: Aaron Dorman, "San Carlos Lake's Dry-Up Is Earliest Ever as Water Levels Plummet," *Casa Grande Dispatch*, April 20, 2021.

195 *pay the installments on its debt*: "White Paper: Understanding the CAP Repayment Obligation," Central Arizona Project, accessed November 2021, https://library.cap-az.com/documents/departments/finance/cap-wp-repayment-obligation-010621.pdf.

195 *accept weaker rights to the water*: "Agriculture and the Central Arizona Project," Central Arizona Project, accessed November 2021, https://www.cap-az.com/finances-of-cap/agriculture-and-cap/.

195 *at the very bottom of the totem pole*: DeEtte Person, "A Matter of Priorities," *Know Your Water News*, a publication of the Central Arizona Project, March 17, 2021.

195 *they'd be able to survive the settlement*: Interviews with Cassy England and Paco Ollerton, September 2021.

195–96 *water shortage on the Colorado for the first time*: Henry Fountain, "In a First, U.S. Declares Shortage on Colorado River, Forcing Water Cuts," *New York Times*, August 16, 2021.

196 *amount of water Arizona could take from the reservoirs*: Brad Poole, "Colorado River Shortage to Hit Pinal County Farmers with $66M Loss," Pinal Central, August 12, 2021.

196 *revenue from the reduced acreage*: Interview with Cassy England, September 2021.

197 *self-contained communities, cities within cities*: Martin Sinderman, "Building a Legacy," *Atlanta Business Chronicle*, June 21, 2010.

197 *on a development called Merrill Ranch*: Minutes from the July 1, 2002, meeting of the Florence Town Council. Accessed via the Town of Florence Public Portal.

198 *"beauty of the northern Sonoran Desert"*: Jack Johnson Company, *Merrill Ranch Planned Unit Development*, prepared for Vanguard Properties, November 7, 2003.

198 *while the housing market was hot*: Melissa Morrison, "A Sudden Oasis, or Just Sprawl? Phoenix Exurb Sharpens Growth Debate," *Washington Post*, June 5, 2000.

198 *"tidal wave" of new construction*: "First Plats OK'd—'Tidal Wave' Coming In," *Florence Reminder*, June 9, 2005.

198 *"the first signs" of the coming community*: Lynn Ducey, "As Valley Suburbs Approach, Historic Florence Wonders If Small-Town Charm Can Survive a Steamroller of Development," *East Valley Tribune*, June 27, 2005.

199 *first private retirement community*: "History: The Original Retirement Community," Sun City Arizona, accessed November 2021, https://suncityaz.org/discover/history/.

199 *self-contained metropolises*: "ULI Development Case Studies: Anthem—Phoenix, Arizona," *Urban Land Institute Case Studies* 33, no. 18 (October–December 2003).

199 *where they could retire*: Catherine Reagor, "Land Grabs, Fraud, and Foreclosures: Arizona's Rich History of Real Estate Fraud," *Arizona Republic*, December 29, 2019.

199 *"they'll have to bring in plenty of water"*: Michael F. Wendland, *The Arizona Project: How a Team of Investigative Reporters Got Revenge on Deadline* (Kansas City, MO: Sheed Andrews and McMeel, 1977).

199 *later killed by a car bomb*: Richard Ruelas, "'God, It Must Be Don': The *Republic* Scrambled to Cover the 1976 Bombing of Reporter Don Bolles," *Arizona Republic*, June 1, 2019.

199 *any subdivision they wanted to build*: Rita Maguire, "Water Laws and Regulations in Arizona," presentation for Maguire Pearce & Storey, February 1, 2018.

199 *moratorium on new pumping activity*: Interview with Kathy Ferris, architect of the 1980 Groundwater Management Act, May 2022.

199 *further steps to restrict new pumping*: Kathleen Ferris and Sarah Porter, *The Elusive Concept of an Assured Water Supply: The Role of CAGRD and Replenishment*, Kyl Center for Water Policy at Arizona State University, fall 2019, 8.

200 *or stored in reservoirs until it was needed*: Interview with Ted Cooke, general manager of the Central Arizona Project, September 2021.

200 *amount of water that was moving through greater Phoenix*: "Plan of Operation: Submitted Draft," Central Arizona Groundwater Replenishment District, November 8, 2004. Available at https://www.cagrd.com/documents/policies/2006-CAGRD-Conservation-Program.pdf

200 *CAGRD would pick up the tab*: Interview with Sarah Porter, March 2021.

200 *for Merrill Ranch's first phase*: Historical satellite images viewable on Google Earth Pro.

200 *Unable to pay back his loans*: Russell Grantham, "Arbitrators: Developer Should Pay $43.6 Million," *Atlanta Journal-Constitution*, March 10, 2011.

201 *Georgia bank that had backed him later collapsed*: Russell Grantham, "Just Like Atlanta, Developer W. Harrison Merrill Rises Again," *Atlanta Journal-Constitution*, December 14, 2009.

201 *site itself slowed to a standstill*: Lisa Nicita and Carl Holcombe, "Home Sales at Florence's Anthem at Merrill Ranch Slowing," *Arizona Republic*, July 14, 2006.

201 *willing to take a gamble on a long commute*: David Van den Berg, "Merrill Ranch Development Ready to Welcome Residents Next Month," *Arizona Republic*, June 17, 2006.

201 *broken off in midsentence*: Sarah Boggan, "Spotlight on Florence: Sales Slump Slows Buildout," *East Valley Tribune*, August 20, 2007.

202 *almost all of them remained unbuilt*: Arizona Water Blueprint, azwaterblueprint.asu.edu. The blueprint map allows a user to view all of the enrolled CAGRD member lands and member service areas. A satellite overlay shows that most of these enrolled subdivisions have still yet to be built.

202 *for the duration of the downturn*: Interview with Sarah Porter, March 2021. Among the largest sites of new development was the town of Buckeye, west of Phoenix, a small village where tens of thousands of potential units are enrolled in the CAGRD. Local officials in Buckeye have said they want to expand their small town of seventy thousand to a city of more than a million residents.

203 *eighteen-hole golf course*: Author's own observations from a visit to Merrill Ranch in September 2021, including visits to several model homes. Several billboards and posters in the lobby of the model home complex advertised both the development's water access and its future expansions.

203 *far beyond the Pinal County farming community*: "U.S. Projections on Drought-Hit Colorado River Grow More Dire; California Likely to Get More Cuts by 2025," Associated Press, September 22, 2021. See also Joanna Allhands, "Lake Mead Could Be in a Tier 2 Shortage by 2023. What's That Mean for Arizona?" *Arizona Republic*, May 20, 2021.

203 *to constrain their water consumption*: "Colorado River Shortage: 2022 Fact Sheet," Arizona Department of Water Resources and the Colorado River Project, August 13, 2021.

204 *rising from $160 a year to $633 a year*: Ferris and Porter, *The Elusive Concept of an Assured Water Supply*, 22.

204 *"Director's review of the Plan"*: "Plan of Operation: Submitted Draft," Central Arizona Groundwater Replenishment District, December 29, 2014, https://morrisoninstitute.asu.edu/sites/default/files/kyl_center_elusive_concept_101619.docx.pdf.

204 *depths exceeding one thousand feet*: Ian James, "'Our Own Survival Is at Stake': Arizona Is Using Up Its Groundwater, Researchers Warn," *Arizona Republic*, May 13, 2021.

205   *a two-mile fissure*: Tanner Clinch, "New Earth Fissure Discovered South of Arizona City," *Eloy Enterprise*, January 25, 2017.

205   *"decades down the road, not tomorrow"*: Aaron Dorman, "ADWR: Future Development in Pinal County Can't Rely on Groundwater," *Casa Grande Dispatch*, July 2, 2021.

206   *others are rippling and mature*: Author's own observations from a visit to the Gila River Indian Community reservation, September 2021.

207   *controls more water rights than the city of Phoenix*: Sharon Udasin, "A Generational Historic Struggle to Regain Our Water," Ensia, May 14, 2021, https://ensia.com/fea tures/water-rights-gila-river-indian-community-native-american-american-west/.

208   *to become general manager*: Interview with Robert Stone, September 2021.

208   *to get their drinking water*: Jovana Brown, "When Our Water Returns: Gila River Indian Community and Diabetes," Native Case Studies, Evergreen University, October 25, 2009. See also Randal Archibold, "Indians' Water Rights Give Hope for Better Health," *New York Times*, August 30, 2008.

208   *there was never enough of it*: Interview with Robert Stone, September 2021.

208   *around half that amount*: Daniel Kraker, "The New Water Czars," *High Country News*, March 15, 2004.

209   *the same 2004 agreement*: "Arizona Water Settlement Act of 2004," US Bureau of Reclamation, accessed November 2021, https://www.usbr.gov/lc/phoenix/AZ100/2000 /az_water_settlement_2004.html.

209   *unreliable deliveries of bottled water*: Justine Calma, "The Navajo Nation Faced Water Shortages for Generations—and Then the Pandemic Hit," *The Verge*, July 6, 2020.

209   *lacked access to safe and clean drinking water*: Calma, "The Navajo Nation Faced Water Shortages for Generations."

209   *move more water for longer*: Interview with Robert Stone, September 2021.

210   *had rejuvenated life on its banks*: Sharon Udasin, "Innovative Partnerships and Exchanges Are Securing the Gila River Indian Community's Water Future," Circle of Blue, May 13, 2021.

210   *that passes through the Central Arizona Project aqueduct*: "Environmental Assessment: CAP Water Option and Lease from the Gila River Indian Community to Apache Junction's Water Utilities Community Facilities District," US Bureau of Reclamation, October 2011.

210   *the balance sheet of CAGRD*: Elizabeth Whitman, "Gila River Indian Community Approves Major Deal to Provide Water to CAP," *Phoenix New Times*, December 5, 2018.

210   *tend his own backyard garden*: Interview with Robert Stone, September 2021.

211 *below the poverty line*: "Gila River Indian Community Primary Care Area (PCA): 2020 Statistical Profile," Arizona Department of Health Services, February 1, 2021.

CHAPTER 7: BAILOUT

215 *buy a house and start a family*: Interview with Sara Langford, March 2021.

216 WE ❤ OUR CHURCH: Author's own impressions from visit to Larchmont-Edgewater in March 2021.

216 *made Sara's jaw drop*: FEMA's Flood Insurance Rate Maps for Larchmont can be viewed at msc.fema.gov. Also from an interview with Gabriella Beale, March 2021.

216 *disclose a home's flood history to potential buyers*: Aaron Applegate, "Buyer Beware: Check If the Property Is in a Flood Zone," *Virginian-Pilot*, January 24, 2015.

216 *cost more than the house itself*: Interviews with Gabriella Beale and Mike Vernon, March 2021.

217 *didn't require flood insurance*: Interview with Sara Langford, March 2021.

218 *the client backed out*: Interview with Gabriella Beale, March 2021.

218 *"or get a kayak"*: This agent asked me not to use his name for fear that speaking about the flood insurance conditions in his area would jeopardize future business.

219 *once they fall they will not go back up*: Interviews with Benjamin Keys, Jesse Keenan, Rachel Cleetus, and Shana Udvardy, March and April 2021.

219 *median length of homeownership is thirteen years*: Nadine Evangelou, "How Long Do Homeowners Stay in Their Homes?" National Association of Realtors, January 8, 2020.

219 *water in the area to rise five feet*: "Sea Level Rise Viewer," National Oceanic and Atmospheric Administration Office for Coastal Management, accessed November 2021, coast.noaa.gov/digitalcoast/tools/slr.

219 *homeowners, insurers, banks, and all levels of government*: Eric Holthaus, "Rising Seas Could Wipe Out $1 Trillion Worth of U.S. Homes and Businesses," Grist, June 18, 2018.

220–21 *drained down into the depression*: "The Chesapeake Bay Bolide Impact: A New View of Coastal Plain Evolution," US Geological Survey Fact Sheet 049-98, accessed November 2021, https://pubs.usgs.gov/fs/fs49-98/.

221 *over thousands of years*: "Chesapeake Bay Impact Crater Adds to Sea Level Rise," Phys.org, April 27, 2014.

221 *ports on the new continent*: "Chesapeake Bay: History & Culture," National Park Service, accessed November 2021, https://www.nps.gov/chba/learn/historyculture/index.htm. See also "Colonial Period: Jamestown Colony," historical overview published by the Mariner's Museum, accessed November 2021, https://www.marinersmuseum.org/sites/micro/cbhf/colonial/col003.html.

221   *economic linchpin of the Middle Atlantic*: Kenneth Lasson, "Part One: Historical Perspective (of the Chesapeake Bay)," from *Chesapeake Bay in Legal Perspective*, US Department of the Interior publication, March 1970.

221   *along the city's western edge*: "Naval Station Norfolk," published by Encyclopedia Virginia, accessed November 2021, https://encyclopediavirginia.org/entries/naval-station-norfolk/.

221   *more than a foot since 1950*: "Relative Sea Level Trend: 8638610 Sewells Point, Virginia," NOAA Tides and Currents, accessed November 2021, https://tidesandcurrents.noaa.gov/sltrends/sltrends_station.shtml?id=8638610.

221   *faster than almost anywhere else in the country*: Jon Loftis et al., "StormSense: A New Integrated Network of IoT Water Level Sensors in the Smart Cities of Hampton Roads, VA," *Marine Technology Society Journal* 52, no. 2 (March 2018): 56–67.

221   *"park" in front of Norfolk*: Dave Mayfield, "Gulf Stream Emerging as Sea Level Rise 'Wild Card' for Hampton Roads," *Virginian-Pilot*, March 14, 2018.

221   *speckled with bug-filled swamps*: Thomas Dahl and Gregory Allord, "History of Wetlands in the Conterminous United States," from *National Water Survey—Wetland Resources*, US Geological Survey publication, 1996.

222   *manipulated the natural environment*: Interview with Troy Valos, special collections consultant at the Norfolk Public Library, March 2021.

222   *what used to be creek beds and marshes*: "Map of the Country Contiguous to Norfolk," US War Department, circa 1812, National Archives. Thank you to Troy Valos for help interpreting this map and several others.

222   *never hard to come by after a bad storm*: Interviews with Troy Valos, Mike Vernon, Orrin Pilkey, and A. R. Siders. See also Gilbert Gaul, *The Geography of Risk* (New York: Sarah Crichton Books, 2019).

222   *and stayed clogged for longer*: Peter Coutu, "In Norfolk, Sea Level Rise Reduces Some Stormwater System Capacity by 50%, Data Shows," *Virginian-Pilot*, January 3, 2021.

222   *shopping centers near the highway*: On the occasion of Norfolk's "Dutch Dialogues" discussion event to discuss sea-level rise, Troy Valos created a digital map overlay that layers recent maps of tidal flooding on top of historical creek bed locations within Norfolk. The maps show an almost one-to-one correspondence between areas that flood today and areas that used to be water or marshland.

223   *the ammunition depots*: Nicholas Kusnetz, "Rising Seas Are Flooding Norfolk Naval Base, and There's No Plan to Fix It," Inside Climate News, October 25, 2017.

223   *move on to Alex's next deployment*: Interview with Alex Lane, April 2021.

224 *a more than threefold increase*: Thank you to Alex and Kezi Lane for allowing me to view this letter.

226 *racking up commissions as he went*: Interview with Mike Vernon, March 2021. I am indebted to Mike for showing me around Norfolk and Virginia Beach, and for his infinite patience when explaining the complexities of flood insurance.

227 *turned risk into certainty*: Brooke Jarvis, "When Rising Seas Transform Risk into Certainty," *New York Times Magazine*, April 18, 2017. I am indebted to Brooke Jarvis for exceptional reporting on real estate and climate adaptation, among other topics.

228 *not high enough to make it unsalable*: Interviews with Mike Vernon, Alex Lane, and Jack Blake, another Norfolk real estate agent, March 2021.

228 *people dissatisfied with the National Flood Insurance Program*: Susan Taylor Martin, "Lloyd's of London Dramatically Lowers Its Flood Insurance Rates in Florida," *Tampa Bay Times*, April 17, 2014.

229 *finance the cost of raising their homes up front*: Interview with Mike Vernon, March 2021. Observations from author's visit to Lynnhaven Colony with Vernon.

230 *near the end of the block*: All observations from author's visit to Chesterfield Heights in March 2021, and from subsequent interviews with Karen Speights over the course of the following months.

231 *Resilience Park*: Interview with George Homewood, planning director for the city of Norfolk, and Doug Beaver, resilience director for the city of Norfolk, March 2021.

231 *absorb the impacts of future disasters*: "What Is Climate Resilience, and Why Does It Matter?" Center for Climate and Energy Solutions, April 2019.

232 *recovery spending after a disaster*: Benjamin Schneider, "Disaster Resilience Saves Six Times as Much as It Costs," *Bloomberg CityLab*, January 17, 2018.

232 *need help around the house*: Interviews with Karen Speights, March and April 2021.

234 *before the autumn high tides*: Interviews with Karen Speights and Kyle Spencer, deputy resilience officer for the city of Norfolk, who has spearheaded the Ohio Creek project, March and April 2021.

234 *the flooding would ever get fixed*: Interviews with Paige Pollard and Mason Andrews, March 2021. The students who created the plan were Andrews's students from Hampton University.

234 *explore solutions to the city's flood problems*: The Dutch embassy helped coordinate a number of these talks in several flood-prone cities, part of an apparent attempt to market Dutch engineering knowledge to waterlogged American cities. The first

was held in New Orleans, with others taking place in New York, Charleston, and St. Louis.

235 *no one thought it would ever get built*: Interviews with Mason Andrews; Skip Stiles, president of the advocacy organization Wetlands Watch; and David Waggoner, principal at Waggoner Ball, the architecture firm that designed the Ohio Creek project, all in March 2021.

235 *same climate issues that Chesterfield Heights did*: Nicholas Kusnetz, "Norfolk Wants to Remake Itself as Sea Level Rises, but Who Will Be Left Behind?" *Inside Climate News*, May 21, 2018. See also "Gardening Tidewaters," episode from the *Broken Ground* podcast, Southern Environmental Law Center, published June 2020.

235 *from walking to school*: Interview with Kim Sudderth, organizer from Mothers Out Front who helped advocate for a bus in Tidewater Gardens, April 2021. Michelle Cook passed away from complications of COVID-19 in late 2020.

235 *biggest public housing projects*: Caleb Melby, "A Virginia City's Playbook for Urban Renewal: Move Out the Poor," *Bloomberg Businessweek*, September 22, 2020. See also Dana Smith, "Over Half of Tidewater Gardens Residents Re-Located to Make Way for Multi-Million Dollar Redevelopment," 13News Now, June 29, 2021.

235 *or just too controversial*: Interview with David Waggoner, architect at Waggoner Ball, and Mason Andrews, whose class at Hampton University designed the original Chesterfield flood resilience plan, March 2021.

237 *increase tenfold by the end of the century*: This data comes from Flood Factor, an interactive risk mapping tool created by First Street Foundation using proprietary satellite data. You can view flood risk for your own community at floodfactor.com.

237 *within an hour's drive of the ocean*: "What Percentage of the American Population Lives Near the Coast?," NOAA National Ocean Service, accessed November 2021, https://oceanservice.noaa.gov/facts/population.html.

237 *between half a trillion and two trillion dollars*: "Underwater: Rising Seas, Chronic Floods, and the Implications for US Coastal Real Estate," report by the Union of Concerned Scientists, June 18, 2018.

237 *an industry like construction*: "GDP by Industry," US Bureau of Economic Analysis, accessed November 2021, https://www.bea.gov/data/gdp/gdp-industry.

238 *which communities the city will protect*: "Norfolk Vision 2100 Plan," adopted by the city on November 22, 2016. See also Eric Hartley, "The Norfolk of the Future Will Move Away from the Waterfront," *Virginian-Pilot*, August 18, 2016. The need to establish a hierarchy of neighborhoods is implied by lines such as: "How could the City deter-

mine which of its many stable, well-kept coastal neighborhoods were to be 'protected' and which the City would 'retreat' from?"

239 *countless other structural doodads*: Jim Morrison, "Climate Change Turns the Tide on Waterfront Living," *Washington Post Magazine*, April 13, 2020.

239 *cover up an obvious eyesore*: Author's own impressions from a visit to Larchmont-Edgewater in March 2021.

239 *patches of yellowed grass*: Thank you to Skip Stiles, president of the advocacy organization Wetlands Watch, for helping me identify and interpret these many signs of tidal flooding during a site visit in March 2021.

240 *what will happen once the blast goes off*: Thank you to Benjamin Keys, Rachel Cleetus, A. R. Siders, and Jesse Keenan for help refining this thought experiment.

242 *the greatest flood risk*: "Case Study: A Bank Evaluates the Impact of Physical Climate Risk to Its Mortgages," S&P Global, May 5, 2022.

243 *worth of toxic mortgages*: Jesse Keenan and Jacob Bradt, "Underwaterwriting: From Theory to Empiricism in Regional Mortgage Markets in the U.S.," *Climatic Change* 162, no. 4 (June 4, 2020): 2043–67.

244 *for a period of days*: "Chapter 1: Sandy and Its Impacts," from *A Stronger, More Resilient New York*, City of New York, June 11, 2013.

244 *in another direction*: Interviews with A. R. Siders, November 2020 through April 2021. I am endlessly grateful to Siders for all her assistance on this book project.

245 *the list of case studies she could pick from was very short*: Anne Siders, "Managed Coastal Retreat: A Legal Handbook on Shifting Development Away from Vulnerable Areas," Columbia Public Law Research Paper No. 14-365, November 6, 2013.

245 *from the Kickapoo River in 1978*: Alexandra Tempus, "Moving On," *The Progressive*, December 1, 2018.

245 *a decade later*: Doug Struck, "How a River Town Relocated, with Climate Lessons for Today," *Christian Science Monitor*, July 15, 2021.

245 *the Isle de Jean Charles community in coastal Louisiana*: In addition to the Isle de Jean Charles relocation project discussed in chapter 4, the federal government has also sponsored or considered sponsoring the relocation of a few Indigenous communities in Alaska, among them the villages of Newtok and Shishmaref.

245 *begun to eat away at roadways*: "Erosion Fight Pits Property vs. Environment," *Cape Cod Times*, March 26, 2021.

245 *fallen into the sea*: Susie Cagle, "Deadly Cliffside Collapse Underscores California's Climate-Fueled Crisis," *The Guardian*, August 7, 2019.

245 *who lived on them for generations*: Harrison Smith, "Tangier Island Is Sinking, Its Pop-

ulation Is Shrinking. And These Guys Want to Make It the Oyster Capital of the East Coast," *Washingtonian Magazine*, March 6, 2016. Two other islands in the bay, Smith Island and Holland Island, are also either close to submerged or already uninhabited.

245 *over the coming decades*: Mario Alejandro Ariza, "As Miami Keeps Building, Rising Seas Deepen Its Social Divide," *Yale Environment 360*, September 29, 2020. This calculation assumes two feet of sea-level rise, which in Miami is all but guaranteed over the next fifty years.

247 *cut their losses and leave*: Nathan Rott, "California Has a New Idea for Homes at Risk from Rising Seas: Buy, Rent, Retreat," NPR, March 21, 2021. The bill in question passed the state legislature but was later vetoed by Governor Gavin Newsom.

247 *move to higher ground*: Alex Harris, "A Town's Pioneering Plan to Fund Retreat from Sea Rise: Have New Development Pitch In," *Miami Herald*, December 22, 2019.

247 *land they want to develop*: A few academics brought this idea up in conversation with me, including Benjamin Keys of the University of Pennsylvania and Shana Udvardy at the Union of Concerned Scientists. See also *Managing the Retreat from Rising Seas—Queens, New York Resilient Edgemere Community Plan*, a report by Georgetown Climate Center that discusses land-swap case studies in New York and California.

247 *the city will clear the land*: Interview with Skip Stiles of Wetlands Watch, March 2021.

247–8 *"it's not totally clear where it should be"*: Interviews with A. R. Siders, November 2020 through April 2021.

CHAPTER 8: WHERE WILL WE GO?

251 *what experts then called "global warming"*: Philip Shabecoff, "Global Warming Has Begun, Expert Tells Senate," *New York Times*, June 24, 1988.

251 *deadline for decisive climate aciton*: Kevin Anderson, "Real Clothes for the Emperor: Facing the Challenges of Climate Change," lecture delivered at Cabot Institute, University of Bristol, November 2012.

251 *difficult work of decarbonization*: Jonathan Watts, "Key COP26 Pledges Could Put World 9% Closer to 1.5C Pathway," *The Guardian*, November 11, 2021.

251 *before the year 2030*: "Climate Change: EU to Cut CO2 Emissions by 55% by 2030," BBC News, April 21, 2021.

252 *domestic coal industry*: Steven Lee Myers, "China's Pledge to Be Carbon Neutral by 2060: What It Means," *New York Times*, September 23, 2020.

252 *significant concessions to the oil and gas industries*: Jim Tankersley, "Biden Signs Expansive Health, Climate and Tax Law," *New York Times*, August 16, 2022.

252   *deep reliance on coal power*: Kieran Mulvaney, "The World Is Still Falling Short of Meeting Its Climate Goals," *National Geographic*, October 26, 2021.

252   *barrels of oil overseas*: Isabelle Gerritsen, "Saudi Arabia Pledges Net Zero by 2060, but No Oil Exit Plan," *Climate Home News*, October 25, 2021.

252   *if they fail to follow through*: Steven Mufson and Annabelle Timsit, "'It Is Not Enough': World Leaders React to COP26 Climate Agreement," *Washington Post*, November 14, 2021. See also Lindsay Maizland, "COP26: Here's What Countries Pledged," Council on Foreign Relations, November 15, 2021.

253   *giants like Amazon and Facebook*: Katherine Dunn, "2020 Was the Year of the 'Net Zero by 2050' Commitment. Will 2021 Be the Year We Get the Details?" *Fortune*, January 1, 2021.

253   *by the next decade*: Brad Plumer and Hiroko Tabuchi, "6 Automakers and 30 Countries Say They'll Phase Out Gasoline Car Sales," *New York Times*, November 10, 2021.

253   *for gas-powered vehicles well before then*: Dyllan Furness, "Low-Carbon Aviation Fuels Are on the Horizon. But for Now, Activists Say We Need to Stay Grounded," *The Guardian*, November 11, 2021.

253   *bottom lines of many bankers and investors*: David Benoit, "Financial System Makes Big Promises on Climate Change at COP26 Summit," *Wall Street Journal*, November 3, 2021.

253   *Arctic National Wildlife Refuge*: Joseph Guzman, "Every Major US Bank Has Now Come Out Against Arctic Drilling," *The Hill*, December 1, 2020.

253   *refusing coverage to new fossil fuel projects*: Alexander Sammon, "The Oil Merchant in the Gray Flannel Suit," *American Prospect*, September 29, 2021.

253   *from future logging*: Lisa Temple and James Song, "The Climate Solution Actually Adding Millions of Tons of CO2 into the Atmosphere," ProPublica, April 29, 2021.

254   *by no means set in stone*: "Temperatures," Climate Action Tracker, accessed November 2021, https://climateactiontracker.org/global/temperatures/.

254   *swallow oceanic islands*: Sabrina Shankman, "Coasts Should Plan for 6.5 Feet Sea Level Rise by 2100 as Precaution, Experts Say," *Inside Climate News*, May 21, 2019.

254   *desert temperatures to lethal levels*: "Climate-Fueled Heat Stress Threatens Worker Productivity," *Deutsche Welle*, June 25, 2021.

254   *by the end of the century*: Abrahm Lustgarten, "The Great Climate Migration Has Begun," *New York Times*, July 23, 2020.

254   *the entire span of the Great Migration*: Abrahm Lustgarten, "How Climate Migration Will Reshape America," *New York Times*, September 15, 2020.

256   *the past decade alone*: "County Information: United States," Internal Displacement

Monitoring Centre, accessed November 2021, https://www.internal-displacement.org/countries/united-states.

256 *experienced a weather disaster*: Sara Kaplan and Andrew Ba Tran, "Nearly 1 in 3 Americans Experienced a Weather Disaster This Summer," *Washington Post*, September 4, 2021.

257 *meeting each other for the first time*: Interviews with Ruth Kmiciek, Michael Orr, and Jessica Di Stefano, June through October 2021.

258 *find a home in the Boise area*: Emily Lowe, "Rebuilding Their Paradise: Camp Fire Survivors Relocate to Treasure Valley," *Idaho Press*, December 7, 2019.

258 *shared her values*: Interview with Jessica Di Stefano, June 2021.

259 *never been touched by fire:* Interview with Michael Orr, June 2021.

260 *might never be rebuilt*: Author's own impressions from visit to Lake Charles in August 2021, and from interviews with Tasha Guidry, Lake Charles activist.

260 *during Hurricane Ida*: Interview with Tasha Guidry, August 2021. See also Carly Berlin, "'It's Just a Vicious Cycle': Evictions, Homelessness Surge in Southwest Louisiana after Hurricanes," *Southerly*, February 17, 2021. Carly's coverage of several disasters in Lake Charles is a feat of reporting and was an invaluable inspiration for me in the process of writing this book.

260 *landscape of blue tarps*: Interview with Tasha Guidry.

260 *wherever they could find them*: John Sutter and Sergio Hernandez, "'Exodus' from Puerto Rico: A Visual Guide," *CNN Investigates*, February 21, 2018.

260 *flight from Puerto Rico*: María Padilla and Nancy Rosado, *Tossed to the Wind: Stories of Hurricane Maria Survivors* (Gainesville: University Press of Florida, 2020).

261 *to work manufacturing jobs*: "Puerto Rican Communities in the Valley," *Our Plural History* project at Springfield Technical Community College, accessed November 2021, http://ourpluralhistory.stcc.edu/recentarrivals/puertoricans.html. Also "History of Hispanics in Buffalo, NY," Buffalo Architecture and History, accessed November 2021, https://buffaloah.com/h/hisp/hisp.html.

261 *settled down in Buffalo and Springfield for good*: Caitlin Dewey, "Hurricane Refugees Liked What They Found in Buffalo—So They're Staying," *Buffalo News*, November 24, 2018; and Greta Jochem, "After the Storm: Hurricane Maria Refugees Putting Down Roots in Northampton," *Daily Hampshire Gazette*, September 25, 2018.

261 *or in the local casinos*: Interviews with Betty Medina Lichtenstein, executive director of Enlace de Familias, and Jose Claudio, chief operating officer at New North Citizens' Council, two organizations that assisted with the relocation of Maria refugees in the Springfield area, March and April 2021.

261 *The gap between cities and small towns will only grow wider*: Christopher Flavelle, "Climate Change Is Bankrupting America's Small Towns," *New York Times*, September 2, 2021.

262 *does not face extreme climate risk*: Mathew Hauer et al., "Sea-Level Rise and Human Migration," *Nature Reviews Earth & Environment* 1, no. 1 (January 2020): 28–39.

262 *those in the Southeast like Atlanta*: Mathew Hauer, "Migration Induced by Sea-Level Rise Could Reshape the US Population Landscape," *Nature Climate Change* 7, no. 5 (April 17, 2017): 321–25.

262 *and a more attractive lifestyle*: Interviews with Mathew Hauer, May 2020 and March 2021. I am grateful to Matt for providing extra context on his fascinating research.

262 *with some other family member*: Interview with Sally Cole, April 2021.

263 *the subtitle said*: Copy of *Times-Picayune* article provided by Sally Cole.

263 *referred to as "illegal immigrants"*: Reeve Hamilton, "The Huddled Masses," *Texas Tribune*, August 30, 2010.

263 *ended up staying in the city*: Tom Dart, "'New Orleans West': Houston Is Home for Many Evacuees 10 Years after Katrina," *The Guardian*, August 25, 2015.

265 *15 percent more than did at the turn of the century*: Darryl Cohen, "About 60.2M Live in Areas Most Vulnerable to Hurricanes," US Census Bureau, July 15, 2019.

265 *temperate cities like Detroit and Minneapolis*: Marie Patino, Aaron Kessler, and Sarah Holder, "More Americans Are Leaving Cities, but Don't Call It an Urban Exodus," *Bloomberg CityLab*, April 26, 2021.

265 *to account for population loss*: "2020 Census: Apportionment of the U.S. House of Representatives," US Census Bureau, April 26, 2021.

266 *even if it were more affordable*: "Redfin Survey: Nearly Half of Americans Who Plan to Move Say Natural Disasters, Extreme Temperatures Factored into Their Decision to Relocate," Redfin, April 5, 2021.

266 *whether to move over the next decade*: Byungdoo Kim et al., "Will I Have to Move Because of Climate Change? Perceived Likelihood of Weather- or Climate-Related Relocation Among the US Public," *Climatic Change* 165, no. 1 (March 2021): 9.

266 *grow crops long term*: Interview conducted by email with Oliver Hulland, October 2021.

267 *caused massive land collapses*: Bob Berwyn, "Massive Permafrost Thaw Documented in Canada, Portends Huge Carbon Release," *Inside Climate News*, February 28, 2017.

267 *once too cold to burn*: Robin Dixon, "Siberia's Wildfires Are Bigger Than All the World's Other Blazes Combined," *Washington Post*, August 11, 2021.

267 *enfolds the entire country*: Chi Xu et al., "Future of the Human Climate Niche," *Proceedings of the National Academy of Sciences* 117, no. 21 (May 2020): 11350–55.

268 *than in the surrounding countryside*: "Explainer: Urban Heat Islands," MIT Climate Portal, climate.mit.edu.

268 *watch games in the summer*: Justin Fox, "It's Gotten Too Hot for Outdoor Baseball in Texas," *Bloomberg Opinion*, October 1, 2019.

268 *access to public parks*: Brad Plumer et al., "How Decades of Racist Housing Policy Left Neighborhoods Sweltering," *New York Times*, August 24, 2020.

268 *difference between life and death*: David Montgomery, "8 Charts on How Americans Use Air Conditioning," *Bloomberg CityLab*, July 10, 2019.

269 *in response to the heat stress*: Mike Thomas, "Chicago's Deadly 1995 Heat Wave: An Oral History," *Chicago Magazine*, June 29, 2015.

269 *dehydration episodes that impair cognitive function*: Caroline J. Smith, "Pediatric Thermoregulation: Considerations in the Face of Global Climate Change," *Nutrients* 11, no. 9 (August 2019): 2010.

269 *when the thermometer gets high*: "Heat-Related Health Dangers for Older Adults Soar During the Summer," National Institutes of Health, June 27, 2018.

269 *increased rates of violent crime*: Christopher Ingraham, "Two New Studies Warn That a Warmer World Will Be a More Violent One," *Washington Post*, July 16, 2019.

269 *for many low-income families*: Maddie Kornfeld, "A Pandemic and Surging Summer Heat Leave Thousands Struggling to Pay Utility Bills," *Inside Climate News*, July 14, 2020.

269 *sudden and extended blackouts*: For a recent example of such a blackout, this one caused by a summer storm, see Michael Gold, "Power Outages Hit Manhattan and Queens as Utilities Face Storm Damage," *New York Times*, August 7, 2020.

270 *a double exposure to high temperatures*: Nicola Lacetera, "Impact of Climate Change on Animal Health and Welfare," *Animal Frontiers* 9, no. 1 (January 2019): 26–31.

270 *twenty gallons of water per day*: "Water Requirements for Beef Cattle," UNL Beef, University of Nebraska–Lincoln, July 2015.

270 *financial threat for livestock companies*: Fiona Kinniburgh et al., *Come Heat and High Water: Climate Risk in the Southeastern U.S. and Texas*. Regional report produced by the Risky Business Project, a climate research initiative chaired by Michael Bloomberg, Tom Steyer, and Hank Paulson. The report finds that poultry and swine producers will face significant economic losses as a result of increased energy demand for cooling.

271 *grow slower and reproduce less*: "How Extended High Heat Disrupts Corn Pollination," Cropwatch, University of Nebraska-Lincoln, August 1, 2011.

271 *over the next decade*: Megan Durisin, "Climate Change Will Cut Corn Yields by a Quarter by 2030, NASA Says," *Bloomberg Green*, November 1, 2021.

271 *spend some amount of time outside*: "Study Finds Extreme Heat Could Threaten

$55.4 Billion Annually in Outdoor Worker Earnings by Midcentury," Union of Concerned Scientists, August 15, 2021.

271 *Pacific Northwest in 2021*: Lauren Gurley, "Farmworkers Endure Brutal Conditions During Historic Heat Wave," *Vice*, June 29, 2021.

271 *often causes other symptoms*: Julia Shipley et al., "Heat Is Killing Workers in the U.S.— and There Are No Federal Rules to Protect Them," NPR, August 17, 2021.

271 *punitive action from labor regulators*: Ariel Wittenberg, "OSHA Targets Heat Threats Heightened by Climate Change," *E&E News*, October 26, 2021.

272 *the largest cattle-producing state*: Ted Genoways, "Fear in a Handful of Dust," *New Republic*, April 2015.

272 *Palm Springs or the Grand Canyon*: See, for instance, Melissa Daniels and Mark Olalde, "'Snowbird Season' to Get a Lot Shorter Due to Climate Change, Threatening Valley Tourism," *Palm Springs Desert Post*, September 11, 2020. See also Katie Rice, "Disney World Faces Challenges with Virtual Reality, Climate Change over Next 50 Years," *Orlando Sentinel*, October 1, 2021.

272 *investment in parched southwestern states*: William Pitts, "A huge amount of Arizona water is being used on memes, selfies and viral videos. Here's why," 12 News, May 9, 2022.

272 *much-damaged refinery in southeast Louisiana*: Erwin Seba, "Phillips 66 to convert storm-hit refinery to oil export terminal," Reuters, November 8, 2021.

273 *a state of permanent underemployment*: Eric Bowen et al., "An Overview of the Coal Economy in Appalachia," West Virginia University, commissioned by the Appalachian Regional Commission, January 2018.

275 *"this new type of refugee"*: Shanelle Loren, "Climate Change Will Destroy Communities. Let's Help Them Move Now," Grist, January 26, 2021.

275 *arrived after Hurricane Maria*: Jeremy Deaton, "Buffalo, NY: Your Climate Refuge," *Nexus Media News*, December 19, 2019.

275 *like the ones Hauer has produced*: Nick Swartsell, "Cincinnati as Climate Change Haven? Some Transplants and City Officials Think So," WVXU, January 26, 2021.

275 *have lived there for generations*: John Sutter, "As people flee climate change on the coasts, this Midwest city is trying to become a safe haven," CNN, April 12, 2021.

275 *too appealing from a climate perspective*: Mario Alejandro Ariza, "Is Climate Change Gentrification Really Happening in Miami?" New Tropic, May 17, 2017.

276 *beneath amber-tinted sludge*: Nic Wirtz, and Kirk Semple, "Guatemala Rescuers Search for Scores of People Buried in Mudslide Caused by Eta," *New York Times*, November 7, 2020.

276 *swept them away to their doom*: Ron Brackett, "Dozens Dead as Eta Triggers Cata-

strophic Flooding, Landslides in Central America," *Weather Channel*, November 5, 2020.

276 *the side of a mountain eroded by rain*: "Passenger Truck Crash Kills at Least 17 in Nicaragua," Associated Press, November 23, 2020.

276 *by broken bridges*: Marlon González, "Eta Brings Heavy Rains, Deadly Mudslides to Honduras," Associated Press, November 4, 2020.

277 *not a decision at all*: Nicole Narea, "Migrants Are Heading North Because Central America Never Recovered from Last Year's Hurricanes," *Vox*, March 22, 2021.

277 *from Central America to the United States*: "Central America: Hurricane Mitch," University of California–Davis, *Migration News* 5, no. 12, December 1998.

278 *"conditions correlated with migration and displacement"*: "Report on the Impact of Climate Change on Migration," White House, October 2021, https://www.whitehouse.gov/wp-content/uploads/2021/10/Report-on-the-Impact-of-Climate-Change-on-Migration.pdf.

278 *prolonged drought, killer sandstorms*: Anthony J. Parolari et al., "Climate, Not Conflict, Explains Extreme Middle East Dust Storm," *Environmental Research Letters* 11, no. 11 (November 2016): 114013.

278 *away from rising sea levels*: Ben Walker, "An Island Nation Turns Away from Climate Migration, Despite Rising Seas," *Inside Climate News*, November 20, 2017. See also Sean Gallagher and Eleanor de Jong, "'One Day We'll Disappear': Tuvalu's Sinking Islands," *The Guardian*, May 16, 2019.

278 *from China into a warming Russia*: Abrahm Lustgarten, "How Russia Wins the Climate Crisis," *New York Times*, December 16, 2020.

278 *triple-digit temperatures for multiple weeks*: David Fickling and Ruth Pollard, "India's Deadly Heatwave Will Soon Be a Global Reality," *Bloomberg*, July 7, 2022.

278 *the province of Guangdong:* Tiffany May, "Extreme Weather Hits China With Massive Floods and Scorching Heat," *New York Times*, June 23, 2022.

279 *"miss the full picture of what's happening on the ground"*: Alexandra Tempus, "Are We Thinking About Climate Migration All Wrong?" *Rolling Stone*, March 14, 2020. I am extraordinarily grateful to Alexandra for her years of work on the climate migration beat.

281 *address the full scale of the risk*: Coral Davenport and Christopher Flavelle, "Infrastructure Bill Makes First Major U.S. Investment in Climate Resilience," *New York Times*, November 6, 2021.

281 *funding for resilient infrastructure*: Ella Israeli, "Increasing Equitable Disaster Relief: Ending Cycles of Displacement for Low-Income Renters," Data For Progress, July 7, 2022.

281   *cannot afford basic flood coverage*: Christopher Flavelle, "The Cost of Insuring Expensive Waterfront Homes Is About to Skyrocket," *New York Times*, September 24, 2021.

281   *account for climate change risk in their portfolios*: "Climate and Natural Disaster Risk Management at the Regulated Entities—Request for Input," Federal Housing Finance Agency, Office of the Director, January 2021. The two largest "regulated entities" are Fannie Mae and Freddie Mac.

282   *serve a quarter of eligible Americans*: "Policy Basics: Federal Rental Assistance," Center for Budget and Policy Priorities, November 15, 2017.

283   *land trusts for future migrants*: Ella Fassler, "Activists Are Sharing Land in Vermont with People Escaping Climate Disaster," *Vice*, June 29, 2021.

283   *construction mandates on high ground*: This is a component of the resilience plan adopted by the city of Norfolk.

283   *that facilitated westward expansion*: Hillary Brown and Daniel Brooks, "How 'Managed Retreat' from Climate Change Could Revitalize Rural America: Revisiting the Homestead Act," *The Conversation*, October 18, 2021.

284   *cannot be returned to their country of origin*: Kate Lyons, "Climate Refugees Can't Be Returned Home, Says Landmark UN Human Rights Ruling," *The Guardian*, January 20, 2020.

Index